The American Assembly, *Columbia University*

THE
NUCLEAR POWER
CONTROVERSY

WITHDRAWN

Prentice-Hall, Inc., *Englewood Cliffs, New Jersey*

A SPECTRUM BOOK

Library of Congress Cataloging in Publication Data
Main entry under title:

THE NUCLEAR POWER CONTROVERSY.

(A Spectrum Book)
At head of title: The American Assembly, Columbia
University.

Background papers for the 50th American Assembly,
held at Arden House, Harriman, New York, in April 1976.

"Arthur W. Murphy, editor."

Includes bibliographical references and index.

1. Atomic power — Congresses.   I. Murphy, Arthur W.
II. American Assembly.

TK9006.N818   621.48'3   76-40017
ISBN   0-13-625582-5
ISBN   0-13-625574-4 pbk.

10 9 8 7 6 5 4 3 2 1

PRENTICE-HALL INTERNATIONAL, INC. (*London*)
PRENTICE-HALL OF AUSTRALIA PTY., LTD. (*Sydney*)
PRENTICE-HALL OF CANADA, LTD. (*Toronto*)
PRENTICE-HALL OF INDIA PRIVATE LIMITED (*New Delhi*)
PRENTICE-HALL OF JAPAN, INC. (*Tokyo*)
PRENTICE-HALL OF SOUTHEAST ASIA PTE., LTD. (*Singapore*)
WHITEHALL BOOKS LIMITED (*Wellington, New Zealand*)

Table of Contents

Preface

America's energy needs will probably continue to grow even in the face of vigorous conservation and more efficient use. One means to help meet these needs is nuclear power. Twenty years ago virtually the entire nation was sure of it, and enthusiasm for the peaceful atom was high. Today there is doubt.

Opponents of nuclear energy express their concern about catastrophic reactor accidents and about the dangers of long-term storage of nuclear waste. They fear sabotage and diversion from peaceful uses to bombs. Thus, "we are not ready for a massive expansion of nuclear power and our society is not ready to live with it, nor is the world ready to live with it," says one opponent, advocating "more restrictive policies on the expansion of civilian nuclear power than what is practised now."

But nuclear power has its champions. Some of them, participants in the Fiftieth American Assembly at Arden House, April 1976, declared in their final report that nuclear power is essential and that "the cost to mankind of not pursuing this option could be tragic." The proponents attempt to dispel fears: "Some people think that today's commercial nuclear reactors can run out of control and explode like a gigantic atomic bomb. . . . This is not possible." Some calculators go on to say, for example, that whoever lives within twenty-five miles of a reactor has one chance in five billion per year of being killed in a nuclear accident and that the chance of being killed by lightning is about 2,500 times greater than by nuclear accident.

The controversy continues, sometimes reasoned and moderate, sometimes both shrill and misinformed. This book was put together by Professor Arthur W. Murphy of the Columbia Law School to help the reader measure the risks and benefits of nuclear power and to show that whatever we conclude about uses of energy, by the end of the century, present structures—political, institutional, and psychological—may be inadequate to do the job. And whatever one may think of the views of the authors of these chapters, it should be understood that they are speaking for themselves and not for The American Assembly, which holds no official position.

Clifford C. Nelson
President
The American Assembly

Arthur W. Murphy

Introduction

The immediate future of nuclear power in the United States is a very hot issue. On the one hand, rapid development of nuclear power (including the breeder reactor) is a major element in the administration's plans to achieve energy independence. On the other hand, a number of groups and individuals have declared their implacable opposition to any nuclear power, including that which now exists. In between these extremes, there are advocates of moderate expansion and slowdown.

The issue is already squarely in the political arena. In California an initiative measure, which if enacted—and held valid—could have eliminated nuclear power in that state, was on the ballot for June 8, 1976, and similar measures are in the offing in other states. Although somewhat more muted, the same fight is going on at the national level. Nor is the controversy limited to the domestic atom; loud outcries have recently been raised against export of nuclear technology to "less stable" countries and even more recently serious proposals have been advanced to force other industrial nations to cooperate with us by cutting off their source of enriched uranium.

In assessing where we are, it might be helpful to look back to the

For more than twenty years, ARTHUR W. MURPHY, *Professor of Law at Columbia University, has studied and written on various aspects of nuclear energy. In 1955–57 he directed a study on "Financial Protection Against Atomic Hazards." From 1961–73 he was a member of the Atomic Safety and Licensing Board of the U.S. Atomic Energy Commission. He has been on New York Bar Association committees on "Science and Law" and "Electric Power and the Environment." Professor Murphy was for ten years a member of the New York State Atomic Development Authority and has been a consultant on environmental legislation to the U.S. Administrative Conference.*

last American Assembly on nuclear power which was held in October 1957, when the power program was in its infancy. The Assembly report makes interesting reading today. Among the findings were that "the United States, unlike many other countries, has today and will have for some years to come, plentiful low-cost fuel supplies." The immediate importance of nuclear technology was, therefore, thought to be for export—particularly to Europe where the fuel picture was much less promising. For Great Britain nuclear power was thought to be a must because "increased imports of oil constitute an unacceptable strain upon [her] international balance of payments [and because] recent developments in the Middle East have brought home the dangers of too great reliance upon oil from that area."

Two other conclusions of the 1957 Assembly are also worthy of note. One was that "the dispersion of atomic power reactors throughout the world is inevitable," and that such dispersion would multiply the "dangers of diversion to weapons use." The second was that the AEC should place "primary emphasis on positive accomplishments in the power field."

To judge by the background papers, there was very little discussion of safety, and little of economics. (It was taken for granted that competitive nuclear power was a long way off.) However, international problems and the proper allocation of roles between government and industry were much on the participants' minds.

In the nineteen years that have passed much has changed. No longer can the United States claim plentiful low-cost, nonnuclear fuel. It too has cause to worry about reliance on foreign oil. The nuclear industry is no longer in its infancy. In the first quarter of 1975, forty-five reactors in commercial operation produced 8.5 percent of the electricity sold in the United States. And although the economics are argued, with skyrocketing costs for "clean" fossil fuel, many people in the electric industry are of the view that nuclear has the edge. Internationally, the picture has changed less, except that proliferation is now an urgent issue and the specter of terrorism has been added to that of war. But it is with respect to safety that the most dramatic change has taken place.

Although there has never been an accident at a nuclear power plant causing injury, the fear of catastrophic accidents, plus the perceived problems of waste-storage and the risks of diversion of nuclear materials —especially plutonium—by terrorists have made nuclear safety one of the burning issues of the day. The foremost "activist" of the day, Ralph Nader, has made a major objective of stopping nuclear power completely. In his view, "the stakes are America and its people. This is a survival issue, pure and simple." Fueled by such hyperbole and supported by many segments of the "media," the opposition to nuclear power has now reached fever pitch. The disagreement between proponents and opponents is not merely one of judgment. They are strongly at odds about the basic facts of safety. The "experts" from industry and the AEC-NRC are not trusted—indeed in the eyes of some, knowledge of the subject seems to be regarded as a disqualifying characteristic. Rarely does a week go by that someone does not call for an "independent" inquiry into safety questions. Both the National Academy of Sciences and the Ford Foundation have commissioned studies of nuclear power.

What, one can well ask, in such a climate do we hope to accomplish by a second American Assembly on nuclear power? Ideally one would hope to find some area of common ground on the facts of safety and economics. Too often the debate is carried on with one side claiming that nuclear power is so unsafe that any alternative is preferable—in Mr. Nader's extravagant phrase, that we would be better off using candles for illumination—and the other side implying that the risk is nonexistent, that the economics are clear, etc.

On the question of safety, agreement between the extremes on anything seems a forlorn hope. People looking at the same data come up with wildly differing conclusions on the likelihood of a serious reactor accident, and the consequences should one occur. On the economics we are somewhat more sanguine that, at least among those who accept the necessity of some growth in the consumption of electricity, we can achieve some measure of agreement.

But even if we cannot reach agreement, it should at least be possible to deescalate the level of the rhetoric—to start a dialogue. In the fury of the debate it is sometimes forgotten that there are people—on both sides—who do not see the question in black and white terms; who recognize that there are risks in going ahead, and in not going ahead. We hope, in these papers, to set out the areas of reasonable disagreement, so as to make possible a discussion of the issues on a somewhat less hysterical plane than that now occupied by many contestants.

The preparation of background papers in an area as controversial as nuclear power poses a significant problem of choice of materials and authors. Obviously all aspects of the issue cannot be covered in detail. We chose those which seemed to us central to the current controversy—that is, those which seem likely to determine the answer to the question whether, and how, nuclear power will be pursued as a major part of the solution to our energy problem: safety, economics, government regulation, government-industry cooperation, and international relations. After making that choice, we had to decide how to present the material. Should we invite champions of opposing views to write the "briefs," or should we attempt to have specialists in each area write, as objectively as they could, a single background piece on his subject. We chose the latter course because we felt it was more likely to produce a coherent picture of the issues to be solved than would a series of chapters advancing competing views. In doing so, we recognized that to some people that choice would be conclusive of bias—it is one of the saddest aspects of the nuclear debate that arguments are more often than not *ad hominem*. However, we hedged our bets somewhat by including one avowedly "antinuclear piece."

Before describing briefly the individual chapters, there are some underlying assumptions which ought to be made clear. One underlying assumption is that we, as a society, will choose to satisfy energy demands. There is, of course, a school of thought that to do so is a mistake; that we would all be better off if we rolled back our energy consumption to an earlier and simpler time—presumably retaining the benefits of medical and agricultural research. However attractive such a prospect might be and whatever one's view of the validity of pleas for nonconsumptive life styles, our assumption is that society will opt for a continuation of a fairly high energy economy. This does not mean that conservation should not be pursued as a major goal or that significant savings cannot be achieved through conservation. Moreover, declining birth rates in the United States may make previous estimates of growth rates much too high. But it seems most unlikely that electric energy demand growth rate will be less than 4 percent and probably it will be at least in the 5 to 6 percent range for the rest of this century. As a practical matter, the additional capacity will have to be coal or nuclear, and probably both.

A second underlying assumption is that there will be no "technological fix" of our energy problem over the next twenty-five years. We should pursue fusion; we must accelerate our research in solar energy but no significant relief will come from any now undeveloped technology. Any choice we make involves risk; we must choose one or the other. There will be no *machina ex deo* by which our dilemma can be avoided.

A third assumption is that while for the United States there does exist an alternative—coal—for satisfaction of our energy needs, the same is probably not true for most of the rest of the world. This means that whatever we decide for ourselves will not be decisive of the course others will pursue, a factor particularly important in the context of the proliferation problem.

Finally, we should point out that we are not attempting to answer the question forever. Our focus is the next twenty-five years during which we must live with our resources pretty much as they are, and our basic technology pretty much as it is. Beyond the year 2000, developments in technology, knowledge, etc., may make present calculations meaningless.

The first chapter by David Bodansky and Fred Schmidt, deals with the technical aspects of reactor safety as well as waste storage and plutonium diversion. The paper is intended as a brief introduction to the technical aspects of nuclear power production and an identification of the chief areas of conflict. These matters are, of course, central to any consideration of nuclear power.

Chapter 2 by Michael Murray deals with the economics of nuclear power and focuses specifically on the comparative advantages and disadvantages of coal-fired and nuclear plants of the kind now being built. The primary emphasis is on base load power generation over the next twenty-five years. Mr. Murray also deals to some extent with the implications for the economics of nuclear power generation on the problems encountered in the so-called "back-end" of the nuclear fuel cycle.

Chapter 3 by Fritz Heimann reviews the alternating periods of optimism and pessimism which have marked the first twenty years of private power generation by nuclear means. Mr. Heimann attempts to identify those things which in his view must be done in order to proceed with the development of nuclear power in the United States. He deals in some detail with the question of whether facilities for fuel enrichment and reprocessing should be built and operated by the government or by private industry.

Chapter 4, which I have undertaken to write myself, is a treatment of the nuclear regulatory process. The primary focus is on the operation of the Nuclear Regulatory Commission although there is some discussion of the relationship among federal agencies and between the federal government and the states. Finally, there is a treatment of the controversial Price-Anderson Act which, although technically not concerned with regulation, remains a significant part of the regulatory picture.

Chapter 5 by John Palfrey is a treatment of the international aspects of the problem with primary emphasis upon the situation of the United States as an actual or potential exporter of technology. Mr. Palfrey treats in detail the problem of our relations with other industrial states as competitors in the export of technology with particular reference to the possibilities of joint and unilateral action to prevent the export of technology being accompanied by nuclear proliferation.

Chapter 6 by George Kistiakowsky is significantly different in kind from the earlier chapters. As noted above, it was commissioned as an essentially antinuclear piece to give balance to the overall discussion. Although some of the views are not what might be expected of the nuclear establishment, the authors of the first five chapters are generally pronuclear in outlook. Dr. Kistiakowsky, on the other hand, has publicly expressed serious doubts about the nuclear effort, at least in the short run. His chapter is not prepared as an answer to the others (indeed, he had not seen them when his was submitted) but as a statement by a man of science—not a specialist in the field—about the things which trouble him.

The picture which emerges from these chapters is, we believe, somewhat different from that which one is apt to glean from the daily press. The contest is seen not as one between certainties—of certain disaster on the one hand or happiness ever after on the other. All agree that the risks and benefits are hard to quantify. Safety emerges as a primary objective, but absolute safety does not exist, and the trade offs between power costs and the "last ounce of prevention" are very hard to measure. In at least one critical area, the need to go forward now with plutonium recycle, the papers show a surprising amount of agreement that such a step should be taken with great caution, if at all.

Perhaps the strongest message that emerges is that our present structures are inadequate. Even if we decide, as an abstract matter, that we should proceed more or less vigorously to pursue the nuclear option, there are many troublesome questions, political, institutional, and psychological, which must be faced. It is not at all clear that we can mobilize the necessary resources to do what is necessary even if we can agree on the ultimate objective. The recent experience with attempts to conserve petroleum must make one doubtful that any unpopular steps will be possible until we are on the brink of disaster.

It seems useful to point out that the nuclear question may be only part of a larger problem of long-term resource management and the management of technology. It is hard to know how we got to this point at this time. Looking back to twenty years ago it seems almost inconceivable that we should have reached the point where the giving up of the nuclear option is a distinct possibility. In part, the opposition to nuclear power appears to be a product of long resentment of arrogance and high-handedness on the part of some members of the nuclear establishment; in part it may be more broadly antiscientific in nature, with nuclear power as the symbol of science in general. Certainly it seems to reflect a disenchantment with the highly touted benefits of technology which many took for granted for so long.

Finally, it seems to reflect a belief in simple answers to complex problems and an unwillingness on the part of many people to face up to the difficult problems of choice with which we must deal. It may turn out that as a society we are simply not able to live with the idea of risk. It may, in the last analysis, be that both the pro- and the anti-nukes will turn out to be wrong and that the society when confronted with the question will make no choice at all. In that event it seems highly likely that we will simply drift along from crisis to crisis—that we will neither live with nuclear power nor reconcile ourselves to living without it.

David Bodansky and
Fred H. Schmidt

1

Safety Aspects
of Nuclear Energy

Introduction

Questions of safety, for ourselves and our descendants, lie at the heart of much of the nuclear debate. That this should be so was possibly inevitable, given the association in the public mind between nuclear power and nuclear weapons. And while there are enormous differences, the association is not entirely inappropriate. For both power and weapons, energy is extracted from relatively small amounts of uranium, and in both cases hazardous radioactive materials are produced in large quantity.

For just such reasons unusual attention was paid to safety from the early days of the civilian nuclear power program. Limits on release of radioactivity to the environment were established well before any serious effort was made to control, for example, automotive pollution, and nuclear power plant design, construction, and operation have been under continuous tight regulation since the program was inaugurated.

If measured by the experience to date, there can be little question that these efforts have been successful. There have been no accidents injurious to the public and there has been negligible environmental

DAVID BODANSKY and FRED H. SCHMIDT *are professors of physics at the University of Washington. Professor Bodansky has held Alfred P. Sloan and Guggenheim Research Fellowships. Professor Schmidt has been an engineer for AT&T and a physicist for the Manhattan Atomic Bomb Project and has held Guggenheim and National Science Fellowships. Both are fellows of the American Physical Society.*

pollution. But still it is a matter of burning debate whether this success has been due more to luck or to skill, and whether, as the nuclear program expands, the past excellent record will be maintained. Further, it is feared that some difficulties have merely been postponed, and that waste disposal problems and terrorist activities will haunt us in the future.

Virtually all deleterious effects of nuclear power are associated with radiation. Therefore the present discussion will begin with a brief consideration of the character of nuclear radiation and its effects on man. Later sections will then take up the main specific issues: the hazards encountered in normal reactor operation, the problems of waste handling and disposal, the dangers of nuclear reactor accidents, and the issue of possible nuclear terrorism.

Health Effects of Radiation

HISTORICAL

Man's conscious, or laboratory, acquaintance with highly ionizing radiation dates from the turn of the century. Roentgen discovered x-rays in 1895, and within the next five years Becquerel and the Curies discovered "new" radiations from uranium ore.

Although the benefits of x-rays for diagnostic purposes were recognized almost immediately, the full recognition of health hazards came more slowly. In consequence, excessive radiation exposures were suffered by radiologists, x-ray technicians, and, in a particularly grim story, radium dial painters. The dangers inherent in such activities were first widely recognized in the 1920s. Serious attempts to establish safety standards date from the late 1920s. These were first directed toward problems of occupational hazards, especially from x-rays, but by the 1950s were expanded to consider problems of general exposure of the population as a whole. Present standards are based upon extensive animal studies, and upon human experience obtained primarily from radiation therapy patients, from atomic bomb victims, and from exposed workers.

In contrast to his relatively short "conscious" history, man's unconscious, or ecological, acquaintance with highly ionizing radiation dates from the beginnings of life. The earth is made of elements born billions of years ago in violent stellar events, and many of the radioactive species created at that time persist on earth to this day. Further, the earth has always been bombarded by particles from continuing cosmic activity, the so-called cosmic rays. Together, these radiations have subjected man and his ancestors to substantial radiation exposures. Perhaps no single facet of the environment in which our biological development has taken place has been more constant than the continuous exposure to radiation—for better or worse.

NORMAL RADIATION EXPOSURE

Units—Radiation doses are measured in units called rem (roentgen-equivalent-man). For many purposes, including the present ones, small radiation doses are of interest and the unit mrem (millirem, one thousandth of a rem) is used instead. These units, rem or mrem, correspond to the total biological effect of a given radiation source, based upon the energy deposited by the radiation per gram of tissue with an adjustment for differences in the effects of the different kinds of rays or particles.

When the radiation comes from radioactive nuclei, as distinct, say, from an x-ray machine, the concept of nuclear half-life is important. Each radioactive species decays at a characteristic rate, denoted by the "half-life." More specifically, the half-life is the time for one-half of a given number of nuclei to decay. Starting with a given number of nuclei, after one half-life one-half are left, after two half-lives one-quarter are left, and so on. After ten half-lives only about 0.1 percent of the activity remains and after twenty half-lives only about one part in one million remains. The half-lives of different substances vary widely, from small fractions of a second for some to billions of years for others.

Natural Sources of Radiation—Man receives radiation doses from the natural environment in three ways: from radioactive elements in the body, from radioactive elements in the earth's crust, and from cosmic rays. The internal radiation gives a dose to the body as a whole of 25 mrem per year, mostly from the isotope potassium-40, which has a half-life of over one billion years, and is unavoidably present as a legacy from the original synthesis of potassium in the universe. The terrestrial radiation, averaged over the country as a whole, gives a dose of 60 mrem per year and the cosmic rays a further 45 mrem per year for an annual over-all total of 130 mrem. This number is an average for the United States, but there are wide variations here and even wider variations if we include other parts of the world. In Colorado, to take an extreme U.S. case, the average terrestrial radiation dose is 105 mrem per year and the average cosmic ray dose is 120 mrem per year, giving a total of 250 mrem per year, 120 mrem per year above the United States average.

Man-made Sources of Radiation—The major exposure from man-made radiation is incurred medically, primarily from diagnostic x-rays and fluoroscopies. These exposures, of course, vary widely from individual to individual. In 1970, the average dose per capita for the entire population was roughly seventy mrem per year. The patients themselves received much higher doses, many thousands of mrem in some cases, but the average is brought down by those who did not receive medical radiation exposures in that year.

EFFECTS OF RADIATION

High Doses—It is abundantly clear that there are serious, harmful effects from large radiation doses, delivered over a short period of time. On the basis of accidents which have occurred and the experiences of the atomic bomb victims, it is known that for a prompt dose of 400,000 mrem or more there is a high chance of death within several days, or at most weeks. Below 150,000 mrem the chief observable acute effects are in the blood count, and below 25,000 mrem there are no medically measurable acute effects.

Low Dose Standards and Studies—Although there are no acute effects, it is believed that there may be long-term medical effects at lower radiation doses, primarily cancer induction and genetic damage.

A comprehensive study of the effects of low radiation doses was prepared in 1972 by the Advisory Committee on the Biological Effects of Ionizing Radiations, under the aegis of the National Academy of Sciences and the National Research Council. The BEIR Report, as it is commonly designated, has been widely accepted as an expression of the best available scientific judgment. Its conclusions are summarized as follows:

Production of Cancer—Conclusions about radiation-induced cancer come primarily from the experience of the atomic bomb victims and of persons who received high doses of radiation in medical treatments. Typically the doses involved were over 50,000 mrem, delivered at a rate of over 1,000 mrem per minute. The chance of cancer-induced death occurring within about twenty-five years after exposure, was deduced to be about 1 or 2 in 100,000 per 100 mrem. For the entire population of the United States, allowing for complexities of age distribution, the BEIR Report estimated that a continual irradiation at the rate of 100 mrem per year corresponds to a cancer toll in the neighborhood of 3,500 per year. This represents about 1 percent of the present cancer rate.

From these numbers it can immediately be understood why no difference in cancer rate between Colorado and the remainder of the

United States can be detected, despite the higher radiation levels in Colorado. The predicted excess cancer deaths for Colorado, with a population of a little over two million, is about forty per year. This is too small an excess to be noted, especially when the other environmental and demographic variables are considered.

Genetic Effects—There is no evidence which directly demonstrates genetic damage from radiation in humans. One might expect that there would be evidence from the exhaustively studied atomic bomb victims. But here the evidence is negative, in that comparing offspring of heavily irradiated parents with offspring of parents who received little or no radiation, no statistically significant differences are seen. In these studies, rates of stillbirths, abnormalities at birth and early deaths among children were all compared with an overall negative result.

Inferences as to genetic effects, therefore, come from animal experiments, primarily with fruit flies and mice. The atomic bomb results only serve to verify that the deduced effects in humans are not thereby substantially *underestimated*.

We can consider the effects of, say, a dose of 3,000 mrem. This could come from an incremental irradiation of 100 mrem per year over a period of thirty years prior to conception, or as a single dose from a medical or accidental exposure. If everyone in the United States were to receive this incremental dose, the BEIR calculations indicate that somewhere between 250 and 4,000 additional genetic defects would appear annually in the next generation, assuming four million births per year (a little high for the early 1970s). Effects would continue in succeeding generations, but at a gradually decreasing rate (assuming no continuation of this increment in the radiation exposure).

These genetic defects would be of the same types already arising from other causes. The normal incidence of genetic defects, where both minor and major ones are included, is over 200,000 per 4 million live births. The postulated radiation dose thus would produce an increase of between about 0.1 percent and 1.6 percent in the present "natural" rate.

THE LINEARITY ASSUMPTION

Virtually all our evidence on the harmful effects of radiation, for studies both in animals and man, comes from observations at high doses. The evidence is of a statistical sort. For example, it is known that for a dose of 100,000 mrem in a brief period, there is about a 1 percent chance of producing cancer within the next thirty years. What is the consequence of a dose of 100 mrem? The "linearity assumption" states that the effect is proportional to the total dose, so that a 100 mrem dose produces a 0.001 percent chance of cancer in thirty years. It also assumes that 1,000 doses of 100 mrem would produce the same effect as a single 100,000 mrem dose. Some support is given to the linearity assumption from the evidence at relatively high doses, but there are also indications that at lower doses at least some effects are less than suggested by linearity.

A contrary assumption is that the body can successfully repair damage from small radiation insults, and that there is a threshold below which no harmful effects occur. The issue of threshold vs linearity could be resolved, or an intermediate response found, either from conclusive experiments at low radiation doses or from a complete understanding of the biological processes by which radiation damage occurs. Neither is available. In particular, the effects postulated at low doses are too small to be established quantitatively in the human and animal populations of the sizes which have been studied.

The BEIR Report adopted the linearity assumption as a means for estimating both cancer and genetic risks, terming this assumption for various conditions as "plausible," "conservative," "prudent," and "the only workable approach." It does this in a spirit of caution, not certainty.

There has been some criticism of the use of the linearity assumption, at least as a means of predicting actual damage. Thus the British Medical Research Council in a 1975 report states that "information is accumulating which suggests that this procedure is bound to overestimate risk." Similarly, Laurison S. Taylor, of the National Council on Radiation Protection and Measurements, termed the linearity no-threshold assumptions as "being made with the maximum credible conservatism so that any calculations using them would result in grossly overestimating or maximizing any possible effects." While many would disagree with the term "grossly overestimating," it appears that there is a consensus, except possibly for plutonium, that if the linearity assumption errs it errs on the side of caution.

On the assumption that no dose of radiation is too low to cause some adverse effects, limits are placed upon artificially created radiation exposures. Since about 1960, the standard established both nationally and internationally has been that the man-made exposure (excluding occupational and medical exposures) to the population should on the average not exceed 170 mrem per year, and for no individual (excluding radiation workers) should it exceed 500 mrem per year. For radiation workers, the occupational limit is substantially higher, amounting to

5,000 mrem per year. Coupled with these limits has been the further proviso that the radiation doses be kept "as low as practicable," (recently changed to "as low as is reasonably achievable") in recognition of the fact that it is wise to incur no excess radiation unless there are compensating benefits. This latter standard ("as low as is reasonably achievable") governs discharges from nuclear power plants.

THE HAZARDS OF PLUTONIUM

Plutonium has received special attention in the press since the spring of 1974, when through the writings of Dr. Arthur Tamplin and Dr. Thomas Cochran attention was focused on its asserted special hazards. It had long been recognized that inhalation of plutonium in small amounts was highly dangerous. (Ingestion of plutonium, in perhaps surprising contrast, is not strikingly hazardous. Plutonium taken into the body with food or drink is almost entirely excreted within a short time, and in consequence it is several thousand times less hazardous to eat a given amount of plutonium than to breathe it.)

The essential thesis of Tamplin and Cochran was that the risks of plutonium inhalation, already believed to be great, had been underestimated by a factor of 115,000. They based this contention on the "hot-particle" hypothesis, namely the hypothesis that plutonium concentrated in small dust particles and inhaled into the lung would produce severe damage in the immediate vicinity of the particle and cause more cancers than the same amount of plutonium distributed uniformly in the lungs.

The hot-particle hypothesis was not a new one, but it had never gained wide acceptance. The writings of Tamplin and Cochran stimulated a reexamination of the problem, and again the weight of authoritative analysis has rejected the hot-particle hypothesis. A special report of the British Medical Research Council in early 1975 concluded that "there is no evidence to suggest that irradiation of the lung by particles of plutonium is likely to be markedly more carcinogenic than when the same activity is uniformly distributed." The hot-particle hypothesis was also reevaluated and rejected in late 1974 in a study by Bair, Richmond, and Wachholz, a trio of American scientists with long experience with plutonium toxicity.

Perhaps the simplest and most readily understood evidence against the hot-particle hypothesis comes from the observation of workers, each in groups of about twenty-five, who suffered over-exposure to plutonium. The first series of exposures were in the atomic bomb program at Los

Alamos in 1944 and 1945 and the second was from a fire at a weapons program plutonium plant at Rocky Flats, Colorado, in 1965. Judged by the hot-particle hypothesis, the Los Alamos victims received doses sufficient to produce a total of 5,000 cancer tumors, but as of 1974 none had been observed. Similarly, the Rocky Flats workers should have eventually accumulated five to fifty cancers each; after about ten years there were no cancers, although some would have been predicted that soon.

While the hot-particle hypothesis has, in our view, been convincingly refuted, the debate concerning plutonium toxicity goes on. For example, it has been suggested, and also vigorously disputed, that smoking increases retention of plutonium in the lung, leading to a high incidence of lung cancer among smokers exposed to plutonium. These arguments persist because the absence of direct evidence of any plutonium damage to man makes estimating the precise degree of hazard a matter of correct interpretation of indirect information.

On two points, however, there is a broad consensus. First, plutonium is a very hazardous substance and inhalation of even small amounts will probably cause cancer. Second, despite this potential hazard and the massive use of plutonium in the weapons industry since 1944, there is no evidence that plutonium workers have suffered ill-effects from plutonium inhalation or ingestion.

Radiation Hazards in Normal Operation

SOURCES OF RADIATION

The basic fuel for a nuclear reactor is enriched uranium contained in many hollow metal cylinders called fuel rods. Cooling water flowing in contact with the individual fuel rods carries away the heat and produces steam used in the electric power generator.

There are two chief sources of radioactive effluents in normal operation. Constituents of the cooling water itself are made radioactive by the large fluxes of neutrons in the reactor. In addition, small leaks in the walls of the fuel rods may allow material, particularly gases, to escape from the fuel rods into the cooling water. This is not as serious as it may sound, because the primary cooling water goes in a closed loop with no direct discharge into the environment. Further, the water and the gases are processed to remove radioactive materials, and potentially hazardous gases are retained until the shorter lived activities have died out. Nevertheless some radioactive emissions will occur, for example from intentional releases of gas after intermediate storage periods and from unintentional small water leaks.

AMOUNT OF RADIATION FROM REACTORS

The extent of the release of radioactive substances from reactors, through mechanisms such as described above, depends upon the engineering care taken to minimize them. Satisfactory practices are enforced by federal licensing and inspection procedures, imposed originally by the Atomic Energy Commission and since its demise by the Nuclear Regulatory Commission. Revised guidelines, put into effect in 1971, and codified in a permanent form in 1975, define the "as low as is reasonably achievable" demands on radioactive emissions in terms of limits on radiation exposure to the public. In essence, the rules limit the radiation dose for anyone continually at the boundary of the reactor site to about ten mrem per year. This includes the consequences of reconcentration of some elements in the food chain. This dose is so small—equivalent for most people to one-tenth the incremental dose of moving to Colorado— that it has little significance in terms of effects on any given individual.

To judge the effect on the population as a whole, one must consider the average dose received throughout the country. The radiation from any one reactor decreases rapidly with distance, but in the future we may have many more reactors. Further, it is necessary to consider the contribution to environmental pollution not only from the reactors themselves, but also from plants which reprocess the used reactor fuel. In a 1972 study, the Environmental Protection Agency analyzed the expected situation up until the year 2000. For that year, when it was projected that the nuclear energy capacity would be equivalent to about 750 large (1,000 megawatts) reactors, the increment in the average annual radiation dose to the population as a whole was estimated to be under 0.5 mrem per person.

CONSEQUENCES AND ALTERNATIVES

An incremental dose of 0.5 mrem per year, which is equivalent to a weekend trip to Colorado, may seem completely insignificant. However, it must be noted that for a population of 300 million (in the year 2000) and accepting the linearity hypothesis, it corresponds to roughly twenty-five additional cancer deaths.

Is this an acceptable number? Compared to an annual (future) cancer toll or perhaps 500,000, it again seems trivial. To the individuals involved, could they be identified, it would not be trivial. It is therefore appropriate to look at alternatives. The most realistic ones are coal power or no power. Coal pollution control techniques have lagged so far behind nuclear emission controls that it is difficult to establish a picture as to what "clean" coal power of the future might imply. However, estimates for coal plants with (future) sulfur dioxide scrubbers and with plants located far from cities, suggest a death rate in the neighborhood of a thousand per year, for the same amount of power generation.

A challenge to this conventional view that for the near future we must rely on nuclear power or coal (barring heavy reliance on oil imports) is implicit in recent articles by Barry Commoner. Writing in the *New Yorker Magazine* in February 1976, he argues:

> . . . it is now evident that some three hundred and twenty-five billion barrels of domestic crude oil are available to us. At the present rate of oil consumption, this amount would take care of the total national demand for oil, without any imports, for fifty or sixty years. There is good reason to believe that in that time nearly all our present reliance on oil could be replaced by energy from our one renewable source—the sun.

Unfortunately, however, the encouraging figure of 325 billion barrels is in striking conflict with the most recent and authoritative estimates of oil resources. Leaning on selected earlier estimates by the U.S. Geological Survey, Commoner failed to take cognizance of their more recent estimate, contained in a comprehensive survey undertaken for the Federal Energy Administration and issued in the spring of 1975 as Circular 725, *Geological Estimates of Undiscovered Recoverable Oil and Gas Resources in the United States.* Here their best estimate is 144 billion barrels of domestic oil. This pessimistic appraisal is shared by a National Academy of Sciences study issued in February 1975, *Mineral Resources and the Environment,* which places the oil resources only some 10 percent higher. Thus, if we accept these analyses, our oil resources are less than one half of those assumed by Commoner. This reduces the relatively comfortable "fifty or sixty years" to about 25 years, which gives little time to implement the large-scale exploitation of solar power let alone more speculative alternatives such as fusion power.

CONCLUDING COMMENTS

Assuming a rapidly developing nuclear power program, the average incremental radiation exposure produced in the United States by the year 2000 will be 0.5 mrem per year—about one-half of 1 percent of the natural radiation exposure. In contrast, other pollutants to which we are exposed, such as sulfur dioxide from the burning of fossil fuels, are in amounts far in excess of anything in prior human experience. It thus seems unreasonable to look at this incremental radiation dose as a cause for serious concern. This view is not unique to nuclear proponents. With very few exceptions, it has been tacitly accepted by nuclear opponents as well, at least since about 1972. For this reason the controversy over nuclear safety, for the most part, has moved away from issues of hazards in normal operation to concentrate on waste storage, reactor accidents and materials diversion.

Handling and Storage of Nuclear Wastes

One aspect of nuclear energy which causes widespread concern is the large amounts of radioactive wastes produced. Not only do these wastes constitute an immediate hazard, but they are believed by many to leave a legacy of danger to future generations. In the present section, some of the characteristics of nuclear wastes will be described, and the methods of handling them considered.

NATURE OF RADIOACTIVE WASTES

Origin—The most important nuclear wastes are those produced directly in the nuclear fuel, the so-called high level wastes. There are other wastes, both the unused residue of uranium mining operations and the materials contaminated outside the reactor fuel assembly itself. Good sense and care (not always shown in the past) are required to cope with all these wastes. However, the high level wastes constitute the most important problem. Therefore, most of the public debate and all our discussion here are directed to them.

In the chain reactions which produce nuclear power, the uranium nuclei are fissioned into two nuclei, each with roughly half the original mass. In addition, uranium may be transformed into heavier elements, such as plutonium by capture of one or more neutrons; the elements thus produced are called the actinide elements. A large fraction of the fission products and all the actinides are unstable, and their decay to stable nuclei constitutes the radioactivity of the used reactor fuel.

Activity of Waste—The nuclei with short half-lives decay quickly and soon constitute no problem, and the nuclei with very long half-lives decay so gradually that the material is always relatively "cool." It is the intermediate cases, say nuclei with half-lives between one year and one million years, that pose the greatest potential problems. The most important nuclei, in terms of waste disposal hazards, are the fission products strontium-90 (29 year half-life) and cesium-137 (30 year half-life), and the actinides plutonium-238, plutonium-239, plutonium-240, and americium-241 (86 year, 24,400 year, 6,600 year, and 433 year half-lives respectively).

A rough measure of the problem is given by the magnitude of the activity—the number of radioactive decays which take place per second. The standard unit of activity is the curie, defined as thirty-seven billion disintegrations per second. It is the activity of one gram of radium. A given number of curies, quoted without any point of reference, conveys little meaning. To give some sense of scale, we will consider the uranium ore mined to fuel a typical large reactor (1,000 megawatts) running continuously for one year, and the wastes produced by such a reactor. The original activity of this uranium and its decay products (just the activity which made Madame Curie famous) is roughly 500 curies. Another point of comparison is the activity of a radioactive source used for cancer therapy; these sources are as large as 10,000 curies.

The spent fuel for each year of reactor operation initially contains about five billion curies when it is first removed from the reactor. After 150 days, this dies down to about 135 million curies, and after 10 years to about 13 million curies, or about 0.3 percent of the original. At this time, about 80 percent of the activity comes from the fission products. The largest contributor is cesium-137, with strontium-90 a close second. In another 600 years, the cesium-137 activity drops to about 3 curies, and its activity and that of the other fission products then constitute a negligible problem. The actinide wastes remain at a very much higher level, or at least would be very much higher were nothing done about them. The special handling of the actinides will be discussed separately below.

Other Characteristics—Another way of looking at the waste problem is to consider the volume of the wastes. Assuming, as all plans call for, that the wastes are solidified, then the volume from the annual output of a reactor will be about seventy cubic feet—about equal to the volume of six standard filing cabinets. Although the magnitude of the activity is extremely large, the volume is extremely small.

A final introductory point concerns the nature of the hazards. If one stood a few feet away from a container filled with solidified waste, the alpha particles, which constitute the main activity of the actinides would represent no hazard at all because they cannot traverse even one foot of air, much less the container wall. The beta particles, the main activity of the fission products, would be stopped by one-eighth inch of iron. The gamma rays from both actinides and fission products are the hardest to stop, but five feet of concrete would reduce their intensity by more than a factor of one million and ten feet of concrete would remove all danger. Thus the wastes would pose no problem if they could be kept intact. The hazards arise only if in some manner they escape from their containers and are inhaled or ingested.

STAGES IN THE HANDLING OF RADIOACTIVE WASTES

In this section, we consider the succession of steps taken in the handling of the wastes from the time they are removed from the reactor until the time at which they have been disposed of permanently. Plans will now be discussed. In the following section, the *contrast* between plans and present reality will be discussed.

Initial Cooling and Transport—When the fuel elements are first removed from the reactor they are too hot, both in the literal thermal sense and the colloquial radioactive sense, to be extensively handled. For that reason they are initially placed in cooling ponds at the reactor site where they are cooled and shielded by water. If they remain in cooling ponds for 150 days, their radioactivity is reduced to less than 3 percent of the original activity as is the heat generation.

After this time, the fuel elements can be shipped to a reprocessing plant. About fifty outgoing truck shipments or ten rail shipments would be required per reactor per year. It is necessary to provide thermal cooling in transit to cope with the continued large heat dissipation. Further, the containers must provide radiation shielding and be able to withstand the impact of any transportation accident. The issues involved are simple, and very satisfactory casks are available although they are bulky. In a typical model, a cask containing half a metric ton of fuel weighs about thirty-five metric tons.

Fuel Reprocessing—The destination of the fuel shipments is the reprocessing plant. The chief purposes of reprocessing are to convert the fuel into a form more suitable for long-term storage and to extract specific elements from the fuel.

Reprocessing entails dissolving the fuel in acid, and passing the liquid through appropriate chemical processes. The elements most worth removing, from an economic standpoint, are uranium and plutonium because both can be used for new fuel for other reactors. In such processing, about 0.5 percent of the plutonium remains in the waste, unseparated, thus contributing to the problems of long-term waste disposal.

If desired, a more complete separation of the plutonium and a separation of other actinides, especially americium, could be undertaken, and these could also be recycled through reactors and be transmuted to fission fragments with shorter half-lives.

During the reprocessing, there will be some relatively small release of radioactive materials into the environment, particularly of noble gases. These releases were included in the estimate cited above of 0.5 mrem per person per year from the normal operation of the nuclear program in the year 2000, and are not a major waste hazard.

Waste Solidification—Within five years after the liquefaction and processing, the wastes are required to be converted into solid form. There are a number of processes which have been shown to produce glasses which are very resistant to erosion in water. Tests at high levels of radioactivity indicate that the integrity of the glass will not be threatened by continued radioactive decays within it.

The solid glass is to be sealed in stainless steel canisters, possibly about one foot in diameter and ten feet long. Ten such canisters would hold the waste generated by one reactor in one year. These canisters are to be transferred from the commercial reprocessing plants to federal repositories within ten years after the original processing. At this time each canister will generate heat at a rate of about three kilowatts. (A typical electric heater operates at about one or two kilowatts and is considered suitable for bedroms.) It is clear that air cooling could suffice to remove the three kilowatt heat output of each waste canister for an array of canisters with moderate spacing.

Surface Storage—If the nuclear industry grows at a rapid rate, and if canisters of the size cited above are employed, then by the year 2010 federal repositories will have received about 75,000 canisters, and will be receiving new ones at rates of up to 10,000 per year. A pedestrian solution would be to put the 75,000 canisters in a square grid, 275 canisters on a side, with each canister contained in its own concrete and steel cask for further protection. There is no apparent reason why such casks could not remain in place indefinitely—or at least as long as Roman aqueduct pillars.

If the canisters were placed twenty feet apart, the whole array would occupy about one square mile. Every ten years, another one or two square miles would be needed. The solidified material, already stripped of most of its plutonium, would be useless to prospective terrorists as a source of weapons material. The radioactive material is in an extremely unwieldy form for dynamiting or stealing. However, were further security against such threats desired, it is quite feasible to arrange the canisters in a more tightly packed and guarded form with more bulky outer protection, and in fact, most surface storage proposals have assumed more protected structures.

Long-Term Storage and Disposal—Despite the apparent practicality of possible surface storage arrangements, virtually all planning has been predicated on the assumption that eventually the wastes would be "disposed of" in a more "final" way. This search for a permanent solution is a natural response to the prospect of wastes accumulating over many centuries, and would also provide an added measure of assurance against the possible vagaries of man-made threats to the disposal sites.

Many long-term disposal schemes have been contemplated. Some, especially disposal into space via rockets, have a futuristic air about them. More down-to-earth plans include deposit in a variety of geological sites, particularly salt beds or rock formations with or without prior solidification. The most durable of plans has been the disposal of solidified waste in salt.

Salt beds are attractive locales for waste disposal because their stable history over millions of years holds the promise of future stability and because vast areas of the United States are underlain with salt beds. In addition the salt itself can provide a natural seal for the solid waste, and salt has a thermal conductivity adequate to carry away the heat generated. The long-term stability of the salt beds is of special importance, because the most plausible way for the wastes to get back into the biosphere is through water. The existence of salt beds which have remained in place for millions of years is evidence of the dryness of the environment if man does not intrude.

The advantages of salt bed disposal were summarized in a National Academy of Sciences-National Research Council Report issued in November 1970. This Report also gave tentative endorsement to the choice of a site at Lyons, Kansas, which the Atomic Energy Commission was considering. It called, however, for further studies of a number of potential problems, including investigation of the "location of previous oil and gas wells" to "determine if these former wells have been adequately plugged to avoid an entrance of water to the salt." Subsequently it was found that just this problem existed. Prior drilling operations had introduced significant water and water damage to the site, and the Atomic Energy Commission abandoned consideration of this site. Other sites have been under subsequent and more careful investigation. A promising, undisturbed salt bed site has been identified in southeastern

New Mexico, but as of early 1976 no commitment had been made to use this site.

Deposit in salt mines probably would in the first instances be in a form which would allow retrieval in the event of difficulties with the site, or a desire to recapture the wastes for useful purposes. In the absence of retrieval, the wastes would remain in the salt mines. The long-term safety of these wastes is discussed below.

THE ADMINISTRATIVE CONFUSION

In the previous section we described a scenario for waste management. Current reality is in sharp contrast to this scenario.

Current Status—At the time of writing, in early 1976, there are no reprocessing plants operating in the U.S. to handle commercial fuel. Instead of the spent fuel proceeding in orderly progression from the reactor to reprocessing, solidification, and storage, it is being accumulated in the cooling ponds of the reactors. If the capacity of these ponds is exhausted, alternate centralized cooling ponds will have to be made available. Furthermore, the conversion of radioactive wastes to glassy solids is not being done on an industrial scale, although the component steps have been tested.

The current low state of reprocessing, the reasons for that state, and the prospects for a turn-around are discussed elsewhere in the papers. For our purposes it is enough to emphasize that the problems seem to be organizational rather than technological.

Mishaps in Waste Handling—In obtaining plutonium for nuclear weapons, initially in the haste of World War II, vast amounts of nuclear waste were generated, processed, and stored. All this went off smoothly with virtually no public concern and no serious mishaps for many years.

This favorable picture was jolted in 1973 when some of these wastes stored in liquid form at Hanford, Washington, escaped from tanks into the surrounding ground. Worse still, the leaks were not noted for several months. No harm was done because, not accidentally, the tanks were sited so that the wastes were well isolated from the water table. Nevertheless, it was an indication of bad failures in monitoring, and a penny-wise attitude in design. Had double-walled tanks been used, as was British practice and later American practice, rather than the single-walled tanks actually employed, then with modest surveillance there would have been no leakage to the outside. The wastes are now being solidified, so the immediate problem is solved, but only after considerable damage to public confidence and perhaps to institutional self-confidence.

A second blow came from the Lyons salt mine fiasco. Whatever the technical merits of the case, and whatever the AEC's own internal safeguards might have been against premature development of the Lyons site, the AEC found itself in the position of seeming advocacy of an unsafe site against ultimately successful public opposition. In an institutional sense, the AEC ended up in severe embarrassment. In consequence, searches for new sites are now being undertaken and described with an approach that appears painfully cautious and tentative.

THE LONG-TERM WASTE PROBLEM

While nuclear critics are also concerned with waste management in the "short" term, their chief focus (and that of many waste management experts) has been on the long-term problems, those purportedly lasting for many thousands of years.

The Basis for Concern—The starting point for most discussions of the long-term problem is the fact that plutonium-239 has a half-life of about 24,000 years. Even 500,000 years later (20 half-lives) a millionth of the activity remains. Thus plutonium-239 will remain a source of radioactive emissions as far in the future as one can meaningfully contemplate. Plutonium-239 is not the only long-lived isotope in the wastes. Also present are americium-241 (433 year half-life), iodine-129 (16 million year half-life), and technetium-99 (210,000 year half-life), as well as other isotopes of plutonium. Thus it is clear that there will be contamination in the nuclear waste from a variety of very long-lived materials; plutonium may not even be the worst offender.

However, it should be recalled that we have a significant amount of potassium-40 (1.3 billion year half-life) in our bodies, and the earth's crust is significantly admixed with uranium-238 (4.5 billion year half-life) and radium-226 (1,600 year half-life). Further, and often forgotten, other hazardous substances, such as lead and arsenic, never decay; they have virtually infinite half-lives, if one wishes to use such terminology. Thus the presence of long-lived contaminants, *per se,* is not necessarily a cause of concern.

Amounts of Material Involved—To assess the magnitude of the long-term problem, say after 600 years, we must know the amounts of radioactive materials which will remain. The time of 600 years is chosen as a convenient, if somewhat arbitrary, period to define what is meant by "long-term." It corresponds to about 20 half-lives of strontium-90 and cesium-137; in this time period their activity, initially very large, becomes negligible. It also corresponds to a period over which we can have some judgment as to the durability of man-made structures, including mines.

A large (1,000 megawatts) reactor, starting with uranium fuel and operating continuously for one year, produces about 300 kilograms of plutonium in an array of isotopes. Following standard practice in reprocessing, 99.5 percent of the plutonium can be removed from the spent fuel for return to other reactors. During the next 600 years some of the plutonium isotopes in the waste decay almost completely, and others are augmented by the decay of heavier elements. The plutonium activity in the waste after 600 years will be about 300 curies, mostly from plutonium-240 (6,600 year half-life) and plutonium-239 (24,400 year half-life). In addition, there will be one curie of iodine-129 and about 500 curies of technetium-99.

More difficult to specify crisply is the amount of americium-241. This depends upon rather technical features, including the duration of plutonium-241 decay before reprocessing, the extent of americium separation (if any) in reprocessing, and upon the type of fuel used in the reactor. We will take, as a working number, a 10,000 curies residue of americium-241 in the waste after 600 years. This number could be reduced by a factor of 100 or 1,000 if americium recovery is undertaken in reprocessing. It could be increased by a factor of about 10 if uranium-plutonium fuel is used instead of pure uranium (again, with no americium recovery).

These projections have all referred to one reactor-year of operation. If one envisages the equivalent of 10,000 reactor-years of operation in the entire period before the year 2000 (with waste to be placed in storage by 2010), the numbers above must be multiplied by 10,000. This would imply that three million curies of plutonium and (very roughly) 100 million curies of americium-241, plus lesser amounts of other isotopes, will remain in the year 2010. (Should plutonium not be separated in reprocessing, not too likely an eventuality in view of the value of the plutonium to other light-water reactors and to breeders, the plutonium content of the wastes would be 600 million curies and the americium content about 500 million curies.)

Those opposed to nuclear power see this accumulation of wastes as a vast time bomb waiting to be released on future generations. They argue that whatever precautions are taken, there can be no assurance that social institutions will last long enough to police the burial sites, or that presently unpredictable geological events will not loosen Pandora's box.

Perspective on the Problem—(NOTE: In our discussion of the long-term waste problem we draw heavily on the studies of Professor Bernard L. Cohen, Professor of Physics at the University of Pittsburgh and a past chairman of the Nuclear Physics Division of the American Physical Society.) The waste will be in solidified form, glassified to resist transfer to the biosphere through erosion by water. It will be in sites supposedly free of water. It will probably be at least about 600 meters underground, so that even were water to come in contact with the waste, and were the glass to leach more rapidly than expected, waste transference to the biosphere would be much delayed.

We will not attempt to quantify these levels of protection. Instead, we will make the modest assumption that the sequestering of waste by man will be done at least as successfully as the sequestering of natural uranium and its daughters, such as radium, has been done by nature. This is not a very high standard, as indicated by the presence of natural uranium and radium in river water and the oceans.

On this basis, a perspective as to hazard can be obtained by comparing the amounts of natural activity in the earth's crust to the amounts in the deposited waste. For the area of the United States there are about 40 trillion kilograms of uranium in the earth's crust down to a depth of 600 meters. This corresponds to 13 billion curies of radioactivity for uranium-238 and for each of its alpha-particle emitting daughters, including radium-226, giving a total of about 100 billion curies. The waste disposal program outlined (that is, the waste from 10,000 reactor-years, after 600 years of cooling) will add 0.1 percent to this and could add many times less were americium recovered and recycled. Further, the uranium and its daughters will be virtually unchanged for a billion years, while the americium-241 activity is reduced by a factor of 3,000 in 5,000 years.

Against this line of argument, it can be objected that the waste is concentrated, while the uranium and radium in the earth are widely dispersed. However, the slowness of leaching of the glassified solid waste and of its travel through the earth should ensure that the dose received by any individual is small. If we accept the linearity assumption, then the same amount of health damage is produced by a small dose to, say, 1,000 people as by 0.1 this dose to 10,000 people. Thus it is irrelevant that the wastes are concentrated. On the other hand, if the linearity assumption is abandoned, then the greater concentration of activity in the wastes is relevant. But then the absolute number of anticipated cancers, from ore or wastes, is reduced; that number already (with the linearity assumption) is small—well under one per century for the wastes, in the analysis discussed below.

It might also be objected that plutonium is uniquely hazardous, and therefore a comparison in terms of the number of curies would underestimate the dangers. However, the most dangerous paths to humans from the buried wastes are through ingestion of soluble compounds, and as noted above, ingestion of plutonium is not particularly hazardous. Indeed, the standards of the International Commission on Radiation Protection indicate that for ingestion radium-226 is more than 100 times more hazardous per curie than plutonium or americium isotopes.

The discussion above used one line of argument to demonstrate that the long-term waste disposal problem has been grossly exaggerated in popular imagination. Analogous conclusions have been demonstrated by means of different comparisons. For example, other analyses have shown that the radiotoxicity of the wastes, after only 300 years, is less than the radiotoxicity of the uranium ore mined to generate the given wastes. This does not "clean up" the earth, because most of the original ore hazard comes from the radium-226 which is still in mill tailings. However, it indicates that the incremental hazard from the stored waste is small.

Professor Cohen has made a comparison of the toxicity of the annual yield of radioactive wastes from 400 reactors (the number required to replace all our present electricity generating capacity by nuclear plants) to the toxicity of some relatively familiar substances. Measured in terms of lethal doses, he concludes that *even after only ten years* of decay, the toxicity of the radioactive wastes for inhalation effects is much less than that of our annual industrial usage of chlorine or ammonia, and, for ingestion effects is less than that of our annual usage of copper or barium. Even more importantly, the radioactive wastes are handled with greater care than are the other hazardous substances, so even were the toxicity the same, the wastes would be likely to cause fewer casualties.

In this same study, Cohen has also estimated the absolute number of cancer deaths which would be caused in various scenarios. We will translate his results to our postulated case of the waste from 10,000 reactor-years of operation after 600 years of storage. In one scenario, intentionally extreme, the wastes are removed from storage, dissolved, and are then directly deposited at random in rivers. So handled, they might ultimately cause about 2,500 deaths. In a somewhat more realistic scenario the wastes are left in place, but after 600 years they are attacked by unanticipated ground water. This would lead to less than one chance in 10,000 of one additional cancer death per year. Were ground water to reach the buried wastes after only ten years, these same calculations indicate that the cancer toll would still be less than one per year.

It might be well, at this point, to recapitulate some of the crucial features of the waste disposal situation: (1) The initial levels of activity are very high, and care must be taken in waste handling. The costs of such care are low, amounting, even in meticulous programs, to not more than about 1 percent of the total electricity cost. (2) The volumes of waste are very small, facilitating their handling. (3) Although reprocessing plants for commercial wastes are not currently in operation, two large plants can be available by the late 1970s, and others are under consideration. (4) Processes to convert wastes into durable glass forms have been demonstrated. (5) There are safe methods of storing solid wastes, above ground or underground. (6) For underground storage, the wastes will be essentially innocuous after about 600 years, and pose little hazard at any time after burial.

In our opinion, the main problems besetting the waste disposal programs are organizational. The waste management authorities note that large scale reprocessing will not begin until about 1980, and the waste need not be deposited, even in interim repositories, until 1990. This may appear to give ample time to search for ever better methods, but too leisurely an approach would not take adequate account of the issue of public confidence.

Reactor Accidents and Safety

BASIS FOR THE DEBATE

The possibility that a commercial reactor might suffer an accident causing deaths or property damage outside the power plant site is perhaps the most worrisome of the issues in the nuclear debate. Since there have been no accidents with public consequences through 1975, estimates of the probability of their occurrence rest upon highly technical and complicated extrapolations. These extrapolations are designed first to predict the accident rates and second to predict the consequences. In this section, we examine such predictions, prefaced by a description of past history and of the nature of possible reactor accidents.

Experience with Light-Water Reactors—With only minor exceptions, all the reactors in use or under construction in the United States for the production of electric power are so-called light-water reactors. At the beginning of 1976, there were about fifty-five in operation, with a cumulative total of about 200 reactor-years of operation. Through this time, no one (no reactor worker or member of the public) has been killed or injured in any accident involving radiation, nor has there been any core melt. (The meaning of core melt is discussed below.) There were 112 U.S. naval ships propelled by nuclear reactors of the light-water type by mid-1975. There have been no accidents, including core melts, in 1,300 reactor-years of operation of these naval reactors.

There are about seventy commercial reactors of diverse designs, many of the light-water type, in foreign countries including the USSR. No radiation-related injuries have been reported.

Past Reactor Accidents—Although there have been no radiation-produced deaths in commercial reactor accidents, and no deaths of any sort outside the plant, there have been some fatal "nonnuclear" industrial accidents at the reactor sites themselves. For example, in West Germany in November 1975, two reactor workers were killed by live steam. The accident was not radiation related and could have happened (and has happened) in any sort of steam generating plant.

Considering all reactors in North America and Western Europe (not just commercial or light-water reactors) through 1975 there were five accidents which did or might have had "nuclear" consequences. In only one of these, at an Army test reactor in Idaho in 1961, were there deaths. In this case three Army technicians died partly from mechanical injuries and partly from radiation, although there was no core melt and no significant radioactivity release outside the reactor area.

Among the remaining four accidents, there have been two core melts; neither caused any large release of radioactivity or any injury to workers or the public. One was in a reactor moderated by heavy-water at Chalk River, Canada in 1952, and the other was in the Fermi reactor in Michigan in 1966.

In the Michigan accident, an experimental breeder reactor, intended as a commercial prototype and therefore relatively small, suffered a partial core melt. It was caused by a metal vane which broke free and partially blocked the flow of liquid sodium coolant. Ironically, the metal vane had been added as a safety measure. Although monitoring equipment warned the operators something was amiss, the warnings were misunderstood. No one received excessive radiation, nor is there any indication that the accident came close to "catastrophic," as some people have contended. The reactor was repaired and continued operation. It was later closed down because it was uneconomical.

The only significant release of radioactivity outside the reactor plant occurred at Windscale, England, in 1957 in a graphite-moderated reactor producing weapons plutonium. The release was sufficient to require clean-up actions and disposal of contaminated milk. No one was known to be injured and property damage was slight. The accident was caused by structural changes in the graphite moderator and thus cannot occur in U.S. commercial reactors.

None of the four accidents described above is directly relevant to the debate over reactor safety, because none was in a light-water reactor of the type in commercial use in the United States and in most of the rest of the world.

A fifth accident, the Browns Ferry fire of March 1975, is directly relevant, having occurred at a large commercial reactor operated by the Tennessee Valley Authority. In this accident, a fire was started (by a candle used to search for air leaks) in trays containing control cables. Some safety features of the reactor were simultaneously disabled, but the operators shut down the reactor with no damage to the core, with no release of radiation, and with no injuries. The economic consequences were large, reportedly in the neighborhood of $100 million, primarily for oil for alternative power plants during the period the reactor plant was being repaired. We will return in a later section to the controversy over the implications of this accident: does it indicate that reactors ride on the brink of disaster or does it indicate that reactors have adequate redundant safety features?

NATURE OF NUCLEAR ACCIDENTS

Posing the Problem—For many technologies a safety record such as that achieved by nuclear power plants would be regarded as adequate evidence that the technology is safe enough. However, this is not true for nuclear power; the reason is simply that *predicted* consequences of a major accident (should one happen) have ranged from large to catastrophic. We will discuss below the estimates of probable consequences which have been made. For now it is sufficient to note that the difficulties of calculations are considerable. Since the major risk of death or injury due to an accident would arise from radiation exposure suffered by victims, the calculation of these consequences depends heavily upon knowledge of the effects of radiation. The effects are particularly uncertain at low dose levels, as discussed earlier, and most of the predicted casualties are the result of low-level radiation. Moreover, determining the probability of a serious failure in a power plant is very difficult because it requires consideration of *simultaneous* failures of individual components. Some of these are completely independent of one another, but others are linked in such a way that one component failure may cause another, or several failures may be produced by the same common cause. These difficulties have led to considerable controversy over the calculations and predictions, but as will be seen below, they have not prevented substantial progress from being made in the attainment of useful estimates.

The Source of the Hazard—Some people are under the impression that today's commercial nuclear reactors can run out of control and explode like a gigantic atomic bomb. Fortunately, this is not possible because the nuclear chain reaction is propagated by relatively slow moving neutrons, and because the fuel is only slightly enriched in highly fissionable uranium-235. In a bomb the reaction is propagated by very high speed neutrons and the fuel is much more fissionable. The reactor is akin to a wood fire; the bomb to a dynamite explosion.

The dangers to life in a hypothetical accident are caused chiefly by the enormous quantities of radioactive isotopes produced in the course of reactor operation. (In normal operation these become the radioactive waste.) As mentioned earlier the half-lives of these products extend over a wide range from a few thousandths of a second up to many years. The longer-lived radioisotopes, those with half-lives in the range of a year or more, slowly build up in intensity as the fuel burns its uranium-235 (and some uranium-238) component.

Any event which can release these radioactive products to the environment can produce a health hazard. Interestingly, the chief starting point for such possible events also lies in the radioactivity because it represents a considerable source of stored energy and heat generation. The energy evolution from the radioactivity (as distinct from fission itself) amounts to about 7 percent of the total reactor power output.

If a reactor is abruptly shut down, the power output of the stored energy drops from 7 percent to about 3.5 percent in about 45 seconds. Some six hours later it is about 1 percent of the original operating power level. Seven percent of the power from a modern 1,000 megawatt-electric power plant (taking into account about 33 percent heat-to-electricity conversion efficiency) is 210 megawatts. In consequence, even though the chain reaction has been arrested, the fuel can melt and in so doing may release radioactive products to the environment.

The Principle of Containment—The objectives of the safety systems are first to prevent the fuel structure from destroying itself, and second to "contain" the radioactive products should these first lines of defense fail. Each of these lines of defense is constructed in depth and has numerous components.

The easiest to understand is the containment system, the first of which is the fuel pellets themselves. They are each about the size of a malted milk tablet and composed of uranium dioxide, a highly compact material with a very high melting point (about 5,000° F). Unless a pellet melts, only a small fraction of the gaseous fission products can escape from it.

The fuel pellets are stacked into long hollow tubes fabricated of a special metal called Zircalloy. The tubes are sealed at both ends but sufficient internal space is available to accommodate the gaseous fission products. Groups of such fuel rods are assembled in bundles with space between each for cooling water and insertion of control rods. The Zircalloy cladding of the fuel rods constitutes the second containment defense line.

The core of a nuclear reactor is composed of many bundles of fuel rods, and the whole assembly is enclosed in the reactor pressure vessel. For a large modern reactor this vessel is some forty feet high, twelve feet in diameter, and with a wall thickness of six or more inches. It is made of a special steel chosen for its exceptional strength and durability at high temperatures. The vessel is carefully tested for flaws by means of x-ray radiography. Its strength is better known and it is thus presumably more reliable than, for example, a girder in a bridge, the probability for sudden failure of which has been found to be exceedingly small.

There are thousands of pressure vessels in modern industrial use from small high pressure gas storage cylinders to great boilers in coal-fired power plants. On the basis of the experience with such vessels, reactor safety experts conclude that the possibility of a catastrophic failure of the pressure vessel is so small that it is not a significant factor in evaluating overall risk.

The reactor pressure vessel constitutes the third line of containment defense. However, there are various penetrations into this vessel such as water coolant pipes, control rods, and radiation monitoring equipment, which present possibilities for breaches even if the vessel itself performs its job.

The reactor vessel is housed in a large containment building, which also contains heat exchangers, pumps, and a multitude of other equipment. The containment building typically is a hemispherically domed structure constructed of steel reinforced concrete with walls three or more feet in thickness. It is the final line of defense. Its purpose is partly physical protection of the reactor, and partly to reduce and delay the release of radioactivity to the environment in case the other defense lines should fail.

Core Melts—The sequence of events by which dangerous quantities of radioactivity can reach the environment must originate with a melting of the fuel core, referred to simply as a "core melt." A core melt would breach the first two containment barriers in a short time—the fuel pellets and the fuel rods. For radioactivity to be released to the environment, the pressure vessel and the containment building must also be breached. For most projected core melts breach of the pressure vessel and containment building will occur eventually, but usually the ultimate release of radioactivity will be small.

A core melt can occur only if the water covering the fuel rods is removed. One way for this to happen is for one or another of the inlet water pipes to rupture so completely that a large section opens up, that is, a "guillotine break." Although the probability of such a guillotine break is very, very small—these are large pipes, several inches thick, and made of special stainless steel—should such a break occur, it would cause the high pressure water in the reactor vessel and ancillary systems immediately to turn into steam. This would produce a rapid "blow-down" in which the cooling water would be expelled from the reactor vessel. (For a chain reaction, the water is essential as a moderator so the "blow-down" immediately stops reactor operation.)

The containment building is designed to withstand the force of such a blow-down, but the core with its residual heat soon becomes vulnerable to melting. The event is referred to as a loss-of-coolant accident, or a LOCA.

Automatic safety systems are designed to cope with a LOCA. The control rods are inserted, a high pressure residual heat removal system starts operating, and when the vessel pressure drops below a prescribed value, an emergency cooling system (the ECCS) operates—initially by passive action; that is, neither electric power nor operator action is required. To continue the emergency cooling, electric power, either on- or off-site, must be provided.

If the emergency core cooling systems also fail so that the core does melt, there are various possible further sequences of events. Considerable quantities of radioactivity will be released through the guillotine break into the containment building. A chemical spray system inside the building operates to remove most of the radioactivity. But the molten core may fall to the bottom of the reactor vessel where it may melt its way through the steel wall and spread out onto the concrete base below.

Though this base is typically about sixteen feet thick, it is likely that the core will melt through it too, and penetrate into the earth below. Calculations indicate that this hot mass could melt its way as much as fifty feet into the earth. Due to its high temperature, approaching ground water would be turned into steam and be kept away from the radioactive mass; but if fissures developed in the earth, some radioactivity could escape to the atmosphere.

Ultimately, the molten mass—consisting of earth, rock, concrete, steel, and fuel—would come to rest, cool, and finally solidify into a glasslike radioactive mass. As it cools down, ground water might reach it; but calculations indicate that its radioactive isotopes could be transported away only exceedingly slowly, due both to the very low leaching rates of the glassy mass and to ion exchange processes with minerals present in the earth.

There are other conceivable courses of events that may follow a LOCA. The core might melt slowly or only partially and come to rest at the bottom of a half water-filled vessel, and present no further danger of a radioactive release. In the worst case, the whole core might fall all at once onto a pool of water in the bottom of the vessel and produce a "steam-explosion." It is conceivable that this explosion could rupture the containment building itself, thus opening the way for a large release of radioactivity to the environment.

Qualitative Consequences of a Release of Radioactivity—If a large quantity of radioactive products is released to the environment, there is a spectrum of possible consequences. One may think of four regions around the power plant. People within the first region may receive sufficient radiation (about 750,000 mrem or more) to suffer acute radiation effects, and almost all would then die within a short time. In the second region (320,000 to 750,000 mrem) the radiation effects will be severe enough to produce varying probabilities of death. A third region includes a region where most people will develop radiation sickness, but will not die except possibly years later from latent effects.

A fourth region exists in which the radiation doses received *may* produce long-term effects, namely latent cancers and genetic defects. The validity of predictions of the consequences for this region rests on the applicability of the linear no-threshold hypothesis of radiation effects.

The area encompassed by each of these regions depends on the magnitude of the release, upon the particular radioactivities present at the time of release, and on the prevailing wind direction, velocity, and other atmospheric conditions. The number of people exposed in each region depends upon the population distribution, and on whether there was sufficient warning time for evacuation or for them to take protective cover.

PREDICTING ACCIDENT PROBABILITIES AND CONSEQUENCES

The starting point for an accident is an initiating failure followed by a series of other failures. Calculation of accident probabilities involves determining the probability for each of these failures, taking into account whether they are independent or can be causally related. The probabilities for each individual stage in the series of failures must be known, based on experience with the equipment involved. The resulting combined rate, which is the rate for that accident, may be too small to be known from experience but still may be calculable.

A simple example will illustrate the principle. Nuclear power plants themselves require electric power to operate. It is supplied either by the plant itself, or if the plant shuts down for one or another reason, by the power-line grid to which it is connected. Either source is called "off-site" power.

Off-site power may fail (with a probability known from experience) and therefore diesel-driven, "on-site" standby electric generators are provided for on-site control power. Two independent standby diesel-electric generator sets are provided so that if one fails to start, the other can carry the load. A diesel engine has a certain probability of failing to start properly, just as sometimes an automobile engine does not start. The probability that one diesel will not start can be measured relatively easily by repeated trials. It is about one chance in one hundred. The probability that neither will start is therefore 0.01 times 0.01, or 0.0001. To measure directly the simultaneous diesel failure rate would be very difficult, and to measure it in conjunction with the (rare) failure of off-site power would be impossible, yet the calculated result is entirely reliable, provided there is no common fault or event which prevents both diesels from starting in a correlated way or affects both off-site power and the diesels.

The last proviso in the illustration above indicates that the accident analyses must include the possibility of common-mode failures as well as of failures which can themselves trigger other failures.

In any electric power plant there are many abnormal events, essentially of a routine nature, which given a certain chain of subsequent events could lead to an accident. There are also numerous safety features and lines of defense which must fail in turn before a true accident occurs.

Each potential accident path thus must be analyzed in detail using known failure rates for the individual components in order to find the combined probability of an accident. The extent of the accident so produced must also be evaluated; "small" accidents are more probable than "large" accidents.

Beyond the prediction of the accident itself lies the prediction of the consequences. Here again, certain actuarial facts are available, for example, weather data which give the probability that in a given location the wind blows (say) east where (perhaps) no people live.

Another example can illustrate the sort of calculation needed. Consider the annual probability that a meteor will kill ten people in the United States. This can be calculated using the probability of impact of meteors of various sizes anywhere on the surface of the earth, the damage produced as a function of size, and the population distribution in the United States. Taken together, one can make a reasonable prediction of the likelihood of ten people being killed by a meteor even though this event may never have occurred in recorded history.

THE EARLY STUDIES

In contrast to the emphasis placed on probabilities in the discussion above, the first studies of reactor safety treated issues of probability very casually. In 1957 the AEC published the results of a study known as WASH-740. It was essentially a "worst-case" consequence study. It did not seriously evaluate probabilities either for mechanical failure or for the ensuing spectrum of consequences, other than to describe the accidents as "highly improbable." At that time there were no commercial plants yet in operation, and most of the design features of the present-day plants were yet to be conceived.

WASH-740 analyzed a reactor considered large at the time—150 megawatts of electricity. The study predicted that a worst-case accident could cause 3,400 fatalities and 43,000 serious illnesses. Since reactors today produce 1,000 to 1,200 megawatts, simple "scaling" suggests consequences about seven times greater. It is not surprising that the results of WASH-740 generate concern.

Public concern was not strongly manifested for over a decade, but eventually the specter of such catastrophic accidents stimulated a great deal of criticism of the AEC and of the nuclear industry. Other AEC reports appeared, some of which were internal or incomplete, and were alleged to have been initially suppressed. In 1973 an extensive report (WASH-1250) was published. It is a large compendium, presenting an informative discussion of the reactor safety systems developed to that time, but it still did not contain a comprehensive study of the probabilities of accidents. In addition, between 1957 and 1974 innumerable scientific papers appeared in appropriate technical journals dealing with the many aspects of reactor safety.

THE REACTOR SAFETY STUDY (RASMUSSEN REPORT)

In 1972, an extensive new study of reactor safety was undertaken, directed by Professor Norman C. Rasmussen of the Massachusetts Institute of Technology. About fifty participants and consultants were involved who were not directly connected with the AEC. They were aided by ten AEC employees. The study took three years, involved seventy man-years of effort, and cost about 4 million dollars. Although the study was sponsored and the expense was borne by the AEC (and its successor, NRC), it is viewed by the sponsors and participants as providing an independent assessment.

In the summer of 1974, the initial results of the study, WASH-1400: *Reactor Safety Study: An Assessment of Accident Risks in U.S. Commercial Nuclear Power Plants,* were made available in draft form. It was left as a draft for about a year for the specific purpose of eliciting comments from a wide spectrum of critics. It was issued in final form in October 1975, incorporating changes stimulated in part by the comments and criticisms received.

Summary Result of WASH-1400—In its summary conclusion, WASH-1400 states:

> The likelihood of reactor accidents is much smaller than that of many non-nuclear accidents having similar consequences. All non-nuclear accidents examined in this study, including fires, explosions, toxic chemical releases, dam failures, airplane crashes, earthquakes, hurricanes and tornadoes, are much more likely to occur and can have consequences comparable to, or larger than, those of nuclear accidents.

This summary result, which places the chances of nuclear accidents in perspective with many other risks with which we are familiar, is very comforting, provided one has confidence in all the ramifications of the calculations, as well as in the expertise and integrity of those who made the study. We will return later to the question of believability.

Some Specific Results of WASH-1400—The Reactor Safety Study (often referred to as the Rasmussen Report, but herein simply as the Safety Study) addressed itself only to current U.S. commercial reactors, of which there are two types: the boiling water (BWR) and the pressurized water (PWR) reactors. Each uses ordinary or light water as a coolant and moderator. The results for the two types are rather similar, and we will continue to refer to them without distinction as light-water reactors. The term "reactor safety," as used here, does not include questions of safety in fuel shipment, radioactive waste disposal, or possibilities of sabotage.

Only core melting can lead to public risks, and thus the discussion which follows concerns only core melts. Since no core melts have occurred with light-water reactors during about 2,000 reactor-years of experience (including commercial and naval reactors, domestic and foreign), the *actuarial* probability of a core melt can be said to be less than about one chance in a thousand for each reactor each year. In contrast, the probability calculated in the Safety Study is much lower; *viz.,* one chance in 20,000 per reactor per year. Said another way, this means one might *expect* a core melt to occur every 20,000 years for each reactor. Thus, for 100 reactors, the maximum number addressed by the Safety Study, one might expect one core melt every 200 years.

Although partial core melts are conceptually possible, the Safety Study treated any core melt as total, a procedure which tends to maximize the consequences of a core melt. However, in contrast with previous opinions, the Safety Study found that most core melts would not lead to serious public consequences. Most core melts are predicted to lead to less than one latent cancer per year in subsequent years, and only about one in a hundred or one in two million reactor-years, is predicted to produce any early fatalities. Thus, even in the eventuality of a core melt, the medical consequences would in most cases be small. There would be some radioactive contamination outside the reactor site, and the cost of cleaning it up would most probably be about $1 million. In addition, there would be the large cost of repairing the reactor, and of replacing the electricity generated by it; the electricity is worth over $100 million per year.

Another important finding was that most releases of radioactivity would be preceded by a warning period of several hours, during which people living close to the nuclear plant could be evacuated. This factor is taken into account in translating a given release of radioactive materials into specific predicted consequences, with the effectiveness and speed of evacuation estimated from past experience with evacuations made in response to such events as chlorine gas releases and hurricanes.

The Safety Study quantified each category of accident into a set of consequences, or risks to the public. As the magnitude of the consequence increases, the probability of its occurrence decreases. For example, to consider an accident in the middle of the scale, it is concluded from the calculations that 1 core melt in 500 would produce 110 early fatalities and 3,000 early illnesses. In addition, it might lead to the occurrence of 460 latent cancers per year, 3,500 thyroid nodules per year, and 60 genetic defects per year, all persisting for about 30 years. Thyroid nodules are easily treated medically. The 60 genetic defects per year are the consequences for the first generation; in subsequent generations further defects would appear, but at a progressively lower rate. There would be no physical damage off-site, but there would be property damage of $3 billion for such things as evacuation and decontamination.

With 100 reactors, an accident of this magnitude or larger is calculated to have one chance in 100,000 of occurring in any given year. The increase in cancer incidence and in genetic defects would be difficult to detect because it would spread over a population of about 10 million people. In a population of this size in the United States, the "natural" rates are about 17,000 fatal cancers per year and 8,000 genetic defects per year. Comparing the consequences of such an accident to the "natural" rates in the entire country, the national cancer toll would be increased by about 0.1 percent and the number of genetic defects by about 0.05 percent.

Apart from the estimates the Safety Study has given of risks, it has played a useful role in identifying weak points in reactor systems. Thus, it was found that an unsuspected kind of failure, different from the main coolant pipe failure referred to earlier, was more likely to cause a core melt. The part subject to failure turned out to be relatively easy to modify, and accordingly the needed change is being incorporated into reactors. This example illustrates why the Safety Study suggests that its findings should not be extended beyond one hundred reactors, since improved safety is very likely to be built into future ones.

Further Estimates of Risks—To obtain further perspective on risks one can focus on the worst possible case, on the probable consequences to the country as a whole, or on the risks for any one individual. Comparisons can also be made between reactor risks and other risks encountered. The Safety Study does all of these.

The "worst-case" occurs when there is a core melt accompanied by maximum release of activity and by weather conditions which deposit the maximum amount of radioactive debris in the most populous locations. This event is calculated to produce about 3,000 early fatalities. If one allows for possible errors in this calculation, the "worst-case" might instead be 1,000 or 10,000 early fatalities. Further, *if* low dose rates are effective in producing cancer, there are predicted to be a 30-year total of 45,000 latent cancers; allowing for possible errors in this estimate, the number might conceivably be as high as about 100,000. These numbers correspond to a rather slow-speed evacuation of the population to beyond 25 miles from the reactor, once warning is given of an impending radiation release. Without evacuation, the calculated number of early fatalities rises from 3,000 to about 6,000. The number of latent cancers is essentially unchanged because they occur among a large but distant population.

Whether one takes the lower or higher end of the scale, these are alarming numbers. However, with 100 reactors, the chances of such an accident are calculated to be only one in 10 million per year. Even if the next 900 reactors will be no safer than the first 100 (which we think is unlikely), with 1,000 reactors the chance of such an accident would be one in 10,000 per *century*.

At this point in a discussion, "Murphy's Law" is certain to be quoted by some people: if things can go wrong, they will go wrong. The "one chance in 10,000 per century" is translated into an expectation that it will happen within the decade. This "Law," however, is not an infallible guide to human events: the Tacoma Narrows Bridge was not filled with people when it collapsed; a 747 has not crashed into a football stadium; two 747s have not collided over Times Square; the Strategic Air Command has not accidentally dropped a hydrogen bomb on Chicago. In short, if something is very unlikely, it usually does *not* happen, our pessimism notwithstanding.

For the country as a whole, including possible accidents of all sizes, the predicted number of early fatalities is 0.003 per year with 100 reactors; the other possible consequences are similarly small. For an individual living within twenty-five miles of a reactor, his or her average chance of early death in a nuclear accident is calculated to be one in 5 billion per year.

If such numbers are believed, or thought to be even very rough approximations to reality, then nuclear reactors are spectacularly safe compared to other societal dangers. An individual who tries to avoid automobile hazards by never driving or walking along or across roads still has a 10,000 times greater chance of being killed by an automobile than by a nuclear reactor accident (again for a person living within twenty-five miles of a reactor). The chance of being killed by lightning is about 2,500 times greater than by a nuclear accident.

Even considering only large scale catastrophies (and leaving out war or famine), other possibilities dwarf the dangers of nuclear accidents. For example, there is about one chance in 1,000 per year that a dam failure in the United States will kill 10,000 people—a catastrophe 10,000 times more probable than the "worst-case" nuclear accident. In the worst case, the failure of a major hydroelectric dam might kill 200,000 people.

The net conclusion of these and similar comparisons is that the risks associated with nuclear power plant accidents are very much smaller than innumerable other risks which society accepts rather casually—including man-caused events such as dam failures or chlorine releases and natural events such as hurricanes and earthquakes.

EVALUATIONS OF THE REACTOR SAFETY STUDY

Critiques of Study Draft—As expected there were numerous reactions to the Safety Study when it appeared in draft form in the summer of 1974. For example, a joint review committee of the Union of Concerned Scientists (UCS) and the Sierra Club published a highly critical assessment of the Safety Study draft. One of their main criticisms involved the "methodology" used in calculating probabilities; according to them the chances of accidents should be increased by a factor of thirty or more. Even were one to accept their position, the accident risk due to nuclear power still would be much less than that for many other risks considered. However, evidence in support of the correctness of the methodology has been provided by results for similar applications to other technologies in England, where accident predictions made years earlier have been shown to agree with later events. Other features of the Safety Study were assailed by the UCS-Sierra Club review; these criticisms have in turn been disputed by others; for example, by reactor experts in Norway who have had many years of experience in issues of reactor safety.

An extensive review of the Safety Study draft was made by a committee appointed by the American Physical Society (APS). This review concluded that it has "not uncovered reasons for substantial short-range (in time) concern regarding risks of accidents in light-water reactors." However, it suggested a number of specific areas where continued development could lead to greater degrees of safety including, for example, the need for continued vigilance in the quality of manufacture and in personnel training.

In addition, the APS review contended that the Safety Study draft should have considered latent cancers from low levels of radiation received far from the accident site. According to the APS study, such a consideration would give a total of many more eventual cancer deaths than calculated in the Safety Study draft. It should be pointed out that their calculation depends upon the validity of the linearity assumption.

Revisions in the Study—In its final version, published in late 1975, the Safety Study took cognizance of the APS criticism (as well as of many others) and included the radiation effects at large distances. The incidence of latent cancers from the resulting low radiation levels was included but, partly on the basis of newer data, at a rate about half of that calculated in the BEIR Report. All results quoted above for the Safety Study are from the final version and have incorporated these changes.

Some changes were not trivial. The possible latent cancers were increased by a factor of seven. However, the probable number of early fatalities was reduced by more than a factor of ten, primarily due to a decrease in the calculated probabilities for small accidents. Other changes were not as signfiicant.

Overall, none of the modifications altered the basic conclusion of the Safety Study draft—namely, that accident risks associated with nuclear reactors are extremely low compared with many other natural and man-made risks.

Browns Ferry Accident—Critics of the Safety Study also point to the Browns Ferry accident, referred to earlier, as illustrating the inevitable incompleteness and unreliability of such studies. The supporters of the study argue that the accident falls within broad classes of accidents considered, and therefore that the occurrence of some such accident is no cause for surprise or alarm.

Whether or not the Browns Ferry accident falls within one of the categories considered in the Safety Study, the latter certainly forecasts some accident of equivalent seriousness. Prior to the Browns Ferry accident there had already been about 200 reactor-years of operation of similar reactors in the United States; and if, as the study concludes, there is one chance in 20,000 of a core melt per reactor-year, it was to be expected that during 200 reactor-years there would have been about one accident coming as close as a 1 percent chance of a core melt. In the actual accident, subsequent analysis concluded that at its most serious point there was a 0.3 percent chance of core melt.

Therefore, unfortunate as the Browns Ferry accident was, it does not discredit the Safety Report. The Report will only be discredited if the number of such "near-misses" becomes more frequent than anticipated. It should also be remembered that each accident, the Browns Ferry one and smaller ones as well, is part of a learning process. The particular design mistakes which contributed to Browns Ferry will not be repeated; for example, newer plants will have noninflammable material in the cable system. This does not mean that there will be no mistakes in the future, but it holds out the plausible hope that there will be progressively fewer.

BREEDER REACTOR SAFETY

Reliance on nuclear fission energy beyond a period of thirty to fifty years depends upon successful deployment of breeder reactors which are capable of realizing the full energy potential of uranium-238. Even though only experimental breeders now exist in the United States, a demonstration breeder is now in the planning stage, so the issue of safety for the projected reactors is already beginning to attract public scrutiny.

Many experiments and calculations are at hand which indicate that, although different in detail, the potential for accidents in breeders is probably not quantitatively different than for the light-water reactors. However, no extensive study of the safety of breeders comparable to the Safety Study has yet been made, and the problem of safe breeder design is in large measure a matter for future study and decision.

TRANSPORTATION ACCIDENTS

There are about one million nuclear shipments nationwide each year. The materials shipped range from waste materials of very low radioactivity to highly radioactive spent fuel elements. The shipments are made under regulations whose stringency depends upon the degree of hazard presented. Nuclear materials have been shipped around the nation in considerable numbers for a period of about thirty years, and the record is excellent. There has not been a single injury or death due to the radioactivity in the shipments. Nevertheless, there is a growing public concern over the possibility of accidents during transportation.

Given the good safety record, as in the case of reactor accidents, we must make theoretical predictions to find out what the potential might be for future transportation accidents. We will only examine the case of spent fuel because this represents by far the greatest hazard.

The shipping containers for spent fuel consist of large steel casks which are designed to withstand a variety of very severe accident situations, as for example, falling many feet onto roadways from railroad bridges or being engulfed in an oil fire.

Calculations, based on actual accident data for trucks and assuming 100,000 truck-miles per year per nuclear plant and 200 plants, show that one can expect about 13 highway accidents per year, with 1 death per year from conventional causes. A few percent of such accidents (or about

one per 5 or 10 years) are expected to be accompanied by a minor release
of radioactivity in which no injuries from radiation would occur. Accidents so severe as to release significant radioactivity are much less likely,
and even then the highly lethal area would not extend more than 100
feet or so from the site. Thus, in these shipments there is more danger
in the conventional hazards of transportation accidents than in the nuclear hazards.

CONCLUDING COMMENTS

Our own reading of the Safety Study and the criticisms has given us
considerable confidence in its detailed conclusions. It does not pretend
to be faultless; for example, it is admitted that some possible accident
paths might have been overlooked, but the probability of such oversight
is very low. The fact that the study was originally sponsored by the AEC
does not, in our minds, compromise the conclusions. It should be noted
also, that another independent study of reactor safety was performed
under commission of the Swedish government and that the resulting
report, published in June 1974, gave an equivalently optimistic assessment of reactor safety.

The confidence of the experts and our acceptance of their conclusions
is based not only on the analyses themselves, but upon past history. The
safety of nuclear reactors has been a matter of concern for many years,
and engineered safety features have been under constant study and modification, safety features now accounting for a large fraction of the capital
costs of reactors. The actual safety record of commercial reactors world-
wide, including (reportedly) the USSR, has been outstanding.

Malicious Acts and Nuclear Safeguards

In previous sections we have considered the hazards associated with
nuclear power, assuming good engineering practice and a conscientious
attempt to minimize the hazards. In the present section we discuss dangers
which may arise if deliberate attempts are made to exploit the harmful
potential of nuclear energy, either on the international scene or domestically. Safeguards against such malicious acts will also be discussed.

INTERNATIONAL

The first and most important source of nuclear danger comes from
the possible additional international proliferation of nuclear weapons.
The subject of proliferation and the problems of international safeguards

are extensively discussed in another chapter, and we will devote our attention only to the domestic aspects.

SABOTAGE OF NUCLEAR PLANTS

It is believed by some people that a nuclear power plant would be a natural target for sabotage. This issue was not addressed by the Safety Study, and is not susceptible to the same sort of quantitative study used in the study's analysis, because any answers will depend on the unpredictable issue of how often sabotage will be attempted. Other portions of the fuel cycle are less likely targets because the possibilities for destructive effects are less.

There are several barriers to sabotage of a nuclear plant. First of all, persons entering the premises must pass through a metal detector like those now installed at passenger ramps at hundreds of airports. The effectiveness of these protective devices, together with inspection of hand luggage, has been demonstrated; hijacking of domestic aircraft has essentially vanished. Moreover, because of the sheer massiveness of the structures, large quantities of explosives would be required to damage the reactor and associated physical equipment. Finally, the containment building itself has thick concrete walls and is normally tightly closed by giant bulkhead steel doors.

Thus, to be successful, a sabotage effort would have to be an attack by an armed band, rather than the work of one or two individuals. The band might be able to do little more than disable the control room, and they would have no guarantee of producing any damage to the reactor. To have any serious hope of greater destruction, the saboteurs would need expert knowledge of the particular reactor or the assistance of one or more of the reactor personnel. Perhaps they could then succeed in doing serious damage, possibly even causing a core melt. The most likely consequences of a melt involve no casualties outside the plant. Thus a very elaborate plan with a large skilled group would be required, and still there is no guarantee of success.

NUCLEAR THEFT AND SAFEGUARDS

Vulnerable Points in Fuel Cycle—The chief possibilities for theft occur during the transportation of the fuel. For light-water reactors in commercial use today, fuel is prepared from uranium enriched to only about 3 percent in uranium-235. It cannot be made to explode, so its transport from a fuel fabrication facility to a nuclear plant presents no special safeguard problem beyond protection against accident damage. It is expensive, but not dangerous. Construction of uranium enrichment facilities is beyond the capability of any conceivable domestic terrorist organization, so their possession of slightly enriched uranium would present no danger.

Once fuel has been in a reactor the products of fission are so radioactive that the fuel cannot be handled except with massive specialized equipment. Upon removal from a reactor (normally after three years of "burning") fuel is stored locally until its activity has died away considerably. It is then shipped to a storage area, or to a reprocessing plant. During this transportation step, the spent fuel is still self-protected by virtue of its residual radioactivity.

The spent fuel is chemically processed in reprocessing plants to recover the unburned uranium as well as the plutonium which has been generated during power production. The energy content of the plutonium is from one-fourth to one-third that of the uranium-235 originally present in the fuel. If the energy content of the plutonium is to be realized, it must be sent to a fuel fabrication facility where it can be incorporated into fuel elements, perhaps integrally mixed with enriched uranium, and thence back to nuclear power plants.

Realistic theft possibilities exist in the fuel cycle only from the time when plutonium is chemically separated in the reprocessing plant until it is put back into a nuclear reactor. The places to guard are therefore: (1) the reprocessing and fabrication facilities; and (2) the transportation vehicles between them and from the fabrication facility to the reactor.

Safeguards Against Theft—Although plutonium is radioactive, it is not dangerously so unless it gets inside the body; and therefore it can be handled with relative ease by trained people. Conceivably, personnel in either of the two facilities, the reprocessing and the fabrication plants, could steal it, perhaps a little at a time, for purposes of terrorism. Despite its low level of activity, sensitive devices are available which can detect if a person is carrying as little as one gram of plutonium. To make off with ten kilograms, one gram at a time, would take one person a lifetime. Furthermore safeguards against internal theft are provided by materials control and accountability. Society has already had a great deal of experience in similar situations: *e.g.*, in precious metal mining or fabrication. Guarding these facilities against internal theft should not be difficult.

Special trucks have been designed, built, and tested. They have been in use for many years for the transport of nuclear materials in the nuclear weapons program. Improvements in the security of these vehicles have been made as experience with them was gained. The trucks are so constructed that penetration is exceedingly difficult. If they are hijacked, they "fail-safe" and cannot be operated. Constant communications are maintained by a special radio system for continuous monitoring of the position of each truck. Each truck is accompanied by tactical escort vehicles which cannot be distinguished from similar ordinary run-of-the-mill vehicles on highways. The trucks themselves are similarly indistinguishable.

It is often contended that safeguards is a new area of concern in which nothing has been accomplished. This is simply not true. For thirty years the AEC has been transporting nuclear materials and weapons, and completed nuclear weapons might be more tempting to terrorists than any reactor fuel. The extensive safeguard system thus developed for military purposes is immediately adaptable to the future needs of the civilian power program. Moreover, the system is a model for demonstration to foreign nations.

There are other measures which have been shown to be technically feasible. For example, plutonium can be diluted with chemicals in such a way as to render subsequent separation extremely difficult.

The degree of security afforded by these safeguard techniques is difficult to quantify beyond the actuarial data. For transportation in the weapons program, about 900,000 miles of service was accumulated by mid-1975 without a diversion attempt. This does not, of course, mean that a successful attempt will not be made in the future. Nevertheless, it seems clear that, apart from an attack by a large band of well-armed terrorists, nuclear materials are already being very well safeguarded.

It is sometimes objected that as nuclear power expands, these measures will become very costly, or impose military-like restrictions on our society. However, even for a future array of 1,000 reactors nationwide, the fuel transportation requirements will be small. Present reactors use about thirty-five metric tons of fuel a year, amounting to about five incoming truck shipments; future breeder reactors may require even smaller shipments because the fuel is more highly enriched. If we assume an average distance of 1,000 miles for each shipment, then for each reactor we require about 5,000 truck-miles of transport per year. If each truck is driven (loaded) a modest 20,000 miles a year, then a fleet of about 250 trucks is needed. This hardly constitutes a serious problem, either from the standpoints of visibility or of cost.

Effects of Atmospheric Dispersal of Plutonium—Plutonium has been characterized as being the most toxic substance known to man. Were this statement true, even the sensitive detection equipment, discussed above, would not provide adequate safeguards against theft by employees of processing plants. The actual toxicity of plutonium was discussed earlier. It is dangerous but thousands of times less toxic than many people believe. It remains to place the toxicity in perspective by making estimates of the actual risk potential if plutonium were dispersed by a terrorist.

Dispersal in the atmosphere is the most likely method a terrorist might employ. However, he faces an immediate technical problem for it is not easy to make plutonium particles of the size needed for them to lodge effectively in the lungs. Scientists at the Battelle Laboratories in Washington State tried very hard before they found suitable techniques to generate such particles for use in animal experiments.

Assuming the terrorist solves the above technical problem, how effective would his dispersal be? Calculations for atmospheric dispersal have been made by Professor Bernard Cohen. His results show that the consequences would be surprisingly small. For example, fifteen grams dispersed without warning in a region of high population density might eventually cause one cancer death. With warning and minor protective measures, about 150 grams ($\frac{1}{3}$ of a pound) would be required to produce one cancer death. These calculations are based on the linearity hypothesis. Poisoning of a city water supply is even less effective because plutonium is thousands of times less toxic when ingested than when inhaled. Thus, dispersal of reactor plutonium does not seem to constitute a genuine threat.

Plutonium Bombs—Whether a terrorist organization would be capable of constructing even a "crude" bomb using stolen reactor plutonium, or pure uranium-235 or uranium-233, is a matter of considerable debate. (Uranium-235 and uranium-233 are important for types of reactors other than light-water reactors.) For reasons of military security many aspects of bomb construction are not widely known, and thus the public debate takes place in a partial factual vacuum. Although there are many knowledgeable persons, only a few have come forth with statements on which to base public judgments. It is therefore easy to devise scenarios lending credence to either side of the question.

To discuss the question at all requires a rudimentary knowledge of how bombs work, and so we begin with this subject. A bomb consists of a geometrical distribution of fissionable material which initially does not form a "critical mass." That is, so many of the neutrons produced by fission leak away that no chain reaction is possible.

A bomb is designed first to alter the geometrical distribution of the material, so that the mass becomes "super-critical." Next, the chain reaction is started by means of an "initiator," the purpose of which is to provide a sudden burst of triggering neutrons. A simple way of generating a super-critical mass is to rapidly force two sub-critical masses together. Another way is literally to squash a solid sphere of material to a higher density. The former is the so-called "gun" mechanism, the latter the "implosion" mechanism. There are other possible geometrical arrangements, such as an initial thin spherical shell, which is crushed radially by specially shaped high explosive material.

The amount of fissionable material constituting a critical mass depends upon its surroundings. For example, the fuel assembly in a light-water reactor is not critical unless it is immersed in water. Similarly, bombs are constructed with suitable surroundings to reflect and conserve neutrons. The critical mass depends very strongly on the purity of the fissile material and on the thickness, disposition, and composition of the surrounding substances. For example, to reduce the critical mass of uranium-235 to a reasonable value, (say) about 20 kilograms, requires a surrounding sphere of very pure metallic nickel weighing about 450 kilograms, or about 1,000 pounds. More mass would be required to produce supercriticality.

Although a good bomb requires an initiator to assure proper triggering at the exact instant of maximum criticality, a few neutrons are always present to start a chain reaction. In particular, reactor plutonium contains a contaminating isotope, plutonium-240, which spontaneously fissions and produces such neutrons.

If bomb parts are assembled slowly, radiation generated becomes lethal at the instant of criticality, and well before appreciable energy is produced. Thus, experiments to determine the size of a critical mass for a particular geometrical arrangement are very dangerous, unless performed by experienced people equipped with specialized apparatus.

The presence of plutonium-240 also imposes a lower limit on the rate of assembly required to produce an efficient explosion; if the rate is too low, the explosion will fizzle and force the assembly apart before much material has fissioned. Since one kilogram of plutonium-240 emits about a million neutrons per second, any one of which could trigger the explosion, assembly must be very rapid to prevent a fizzle.

In 1944 the discovery of plutonium-240 in reactor plutonium, together with its spontaneous fission properties, almost caused a cancellation of the entire plutonium bomb portion of the Manhattan Project. At Los

Alamos, New Mexico, a crash program was then instituted to develop the implosion-type bomb, and at Hanford, Washington, the "piles" were operated specially to minimize plutonium-240 production. The difficulties encountered have been described in the official history of the project, as well as in more popular versions, in terms which would strongly discourage a potential bomb maker from pursuing the implosion method.

On the other hand, the gun-type method of assembly, used for the uranium-235 bomb during World War II is much simpler; and it may possibly work well enough, even with reactor plutonium, to produce a "crude" bomb: *i.e.,* one having a poor efficiency yet exploding with the equivalent of 100 tons or more of high explosive. The Hiroshima bomb was equivalent to about 20,000 tons of high explosive; a 100 ton equivalent bomb would, of course, still be exceedingly powerful and destructive.

Plutonium produced in commercial reactors contains from 20 to 30 percent plutonium-240. Actual weapons grade plutonium probably contains much less, perhaps only a few percent, although the United States weapons program may have developed techniques to use less pure plutonium.

A well-trained nuclear scientist can calculate the results, including the uncertainties, to be expected for various rates-of-assembly and various bomb materials and compositions. For example, a report by Swedish scientists shows that for two sub-critical pieces of uranium-235 an assembly speed of a few meters per second would suffice to produce an inefficient bomb; whereas, for commercial reactor plutonium, one-half of the material would have to be propelled into the other half at a speed of about 1,000 meters per second (three times the speed of sound) to produce a crude bomb. To achieve this speed would require a good deal of skill in construction, or modification, of a "gun," and powerful high explosives.

Assessment of the Degree of Difficulty Faced by Bomb Makers—It is clear from the discussion above, that fashioning a bomb from stolen plutonium is not something which could be accomplished by the ordinary terrorist. The individual fanatic, or a group of terrorists without very special skills, would have no chance of success. The scale of effort and expertise needed, however, remains a matter of some debate and uncertainty.

This problem concerns all countries which have nuclear power, not just countries with nuclear weapons. It is therefore not surprising that Swedish physicists have assessed the possibilities of secret bomb construction. A report translated from Swedish concludes with the following statement:

> For an illegal organization, a supply of qualified personnel and (the obtaining of) a supply of fissionable material would entail difficulties. Above all, the recruiting of some 50 men among whom 10 or so are qualified specialists and several years of activity in secrecy could probably not be carried out. These problems combined with the high costs for such an organization make it improbable that the organization would engage in the production of nuclear explosives. A single madman (a terrorist?) completely lacks the possibilities for producing a nuclear explosive.

At the other end of the spectrum of opinions, Theodore B. Taylor, a former bomb designer at Los Alamos, N.M., states in the *Annual Review of Nuclear Science* (1975):

> Conceivably, one person who possessed about one normal density "fast" critical mass of fissionable material and a substantial amount of chemical high explosive could design and build a crude fission bomb.

Still another bomb expert, J. Carson Mark, formerly head of the Los Alamos theoretical division, has stated in a letter to the authors that he believes it would require six highly qualified people:

> If one thinks of a small group wanting to build a bomb, and if one supposes that their primary requirement is that it give a "nuclear yield" (as to say, for example, 'the yield must be at least so much; but it is all right if it should turn out to be a few times larger') then I think that such a device could be designed and built by a group of something like six well-educated people, having competence in as many different fields. As a possible listing of these, one could consider: a chemist or chemical engineer; a nuclear or theoretical physicist; someone able to formulate and carry out complicated calculations, probably requiring the use of a digital computer, on neutronic and hydrodynamic problems; a person familiar with explosives; similarly for electronics; and a mechanically-skilled individual. Among the above (possibly the chemist or physicist) should be one able to attend to the practical problems of health physics which would arise. Clearly, depending on the breadth of experience and competence of the particular individuals involved, the fields of specialization, and even the number of persons, could be varied, so long as areas such as those indicated were covered.

The quotation by Taylor, suggesting that one person might be able to build a bomb, is hard to evaluate because it is not clear whether he has in mind a bomb made of uranium-235 or of plutonium-239, with or without a contaminant of plutonium-240.

The statement by Mark is also unclear as to the type of bomb he has in mind, but the need to assemble a group of six diversely expert people would make the operation very different in scope from a one person effort. The Swedish report and Mark are in agreement on the need for trained people, calling for ten and six, respectively. (The Swedish discussion applied to both plutonium and uranium-235 bombs.) Perhaps it is reasonable to conjecture that the "one person" in the Taylor estimate would have to be a person very much like Taylor himself, in view of his exceptional experience.

In our own view, a lone terrorist is incapable of making a bomb. A plutonium bomb is not a simple device, and even a trained group of terrorists face an array of difficult problems. Just concealing their activities would be difficult during the period while they are trying to make a bomb after the hypothetical theft, because the construction of such devices requires a substantial array of specialized equipment and facilities.

With such a wide range of opinions, one is left with an uneasy sense of uncertainty. It is well, however, to keep in mind that whether it is or is not possible for a group of terrorists to build a bomb becomes moot if the safeguards are essentially impregnable. The facts, as presented, suggest to us that only a very large determined attack by a substantial, well-organized, and heavily armed group could steal and successfully conceal a nuclear shipment.

It may be that the greatest threat lies in uncertainty and fear. If the level of public apprehension is great enough, a terrorist group might achieve its purpose by threatening to disperse plutonium or detonate a plutonium bomb, whether or not they had an effective device, or any device at all.

The cost of a strong safeguards program, both for reactors and against theft, is low. Society can well afford to pay for the insurance this will provide. Of course, no safeguard system can be absolutely perfect. It is possible to imagine scenarios involving inside collusion or outside terrorist coups carried out with the virtuosity of Mission Impossible. But after these possibilities are conceded in the abstract, one must ask why a terrorist group would single out such difficult targets as our domestic nuclear power facilities and materials.

Alternative nuclear targets include the stockpiles of nuclear bombs throughout the world and the materials used in civilian power programs in other countries. Further, there are innumerable nonnuclear options for terrorists, employing chemical or biological agents or conventional explosives. The most modest of imaginations can create scenarios of great horror. Perhaps the most surprising feature, in our present psychological climate, is how few of the potential atrocities are actually attempted.

Summary

Nuclear energy has been studied more carefully than any other commercial or industrial activity of which we are aware. No comparable technology has been introduced with so little injury to the public or to the workers themselves. Highly responsible analyses indicate that this excellent safety record can be maintained in the future.

Nevertheless, many conscientious people believe that the use of nuclear energy entails very serious dangers. In the preceding sections we have considered the primary safety issues. It may be helpful to recapitulate, considering key issues in a rough order of increasing importance.

Perhaps the simplest issue to analyze, and the one on which there is the least controversy, is the possibility of radioactive pollution from normally operating nuclear power plants. The emission standards for nuclear reactors are strict, and they can be met and verified by straightforward procedures.

A second matter, that of very long-term waste disposal, has been the subject of a great deal of (in our view, unwarranted) concern. It is widely asserted that there will be hazards persisting from nuclear wastes for thousands of years, or tens of thousands of years, or even longer. However, if one computes the *amount* of activity in the wastes, it can be seen that there is no significant problem beyond about 600 years. Even after 300 years, the radiotoxicity of the wastes will be less than that of the original ore.

The issue of shorter term waste disposal is less clear cut because very large amounts of radioactivity are present. There is enough experience in the handling of military nuclear wastes and early commercial wastes and from tests of new procedures to lend confidence to the conclusion that the wastes can be safely processed, solidified, and placed into surface or underground storage. Nevertheless, this has been an area where engineering and large-scale development have lagged behind needs. As of the beginning of 1976, neither reprocessing plants nor waste solidification facilities were in operation to handle the commercial wastes which were being generated.

Turning to the question of the safety of commercial reactors, two facts stand out clearly: first, the record to date of deaths or injuries from radiation-related accidents has been impeccable. Second, the most comprehensive study of the problem, the Reactor Safety Study (or Rasmussen Report) concludes that reactors are exceedingly safe.

Although we find the Safety Study to be convincing, and its conclusions compelling, it must be recognized that there is significant dissent. To some extent, the competing judgments about the Safety Study and the reactor accident issue as a whole rely on assessments of plausibility, rather than on rigorous scientific argument. There is no way to prove or disprove the contention that something has been forgotten.

We expect that it will take decades of safe operation before all the doubts wither. In the meantime, perhaps some consolation can be taken from the fact that even in the extremely unlikely event of a major accident, it does not seem likely to be a catastrophe of unparalleled dimensions.

A great deal of attention has been paid to the dangers of plutonium, which may be used as a supplementary fuel in light-water reactors and as a crucial fuel for most projected breeder reactors. Plutonium is clearly a very hazardous substance, but the claims as to plutonium toxicity, especially those based on the "hot-particle" hypothesis, are not supported. Again, if one turns to experience, there is no evidence of any serious injury from the extensive handling of plutonium in the weapons program.

Finally there is the question of nuclear terrorism or sabotage, an issue in part connected with that of plutonium toxicity. Any terrorist attempts must overcome very great obstacles in penetrating the extensive domestic safeguards and in solving the technical problems. These obstacles are formidable, and the prospects of doing even modest damage are uncertain. We believe that domestic terrorists can find many targets in our society which are more attractive.

Finally, we wish to note that this article has addressed only a narrow part of the energy problem. Issues of nuclear safety, although deserving extended consideration, cannot be fully resolved without assessment of the benefits of nuclear energy and of the risks to the country's future were this energy source excluded.

R. Michael Murray, Jr.

2

The Economics
of Electric Power Generation — 1975-2000

In this chapter the comparative economics of those alternative forms of base load electric power generation which are likely to make significant contributions to the nation's generation capacity additions up to the year 2000 will be discussed. This ambitious sounding statement actually reduces to a simple question: which produces the most economic form of electricity, coal-fired boiler or nuclear light-water reactor power generation?

The chapter has five sections:

— The first two—Problem Definition and Background—are intended to provide perspective on the problem.
— The third—Economics of Nuclear Power Generation—describes the nuclear fuel cycle and the costs associated with generating electricity from nuclear power plants.
— The fourth—Economics of Coal Power Generation—discusses the coal cycle and the costs of generating electricity from coal-fired boilers.
— The fifth—Comparison and Conclusions—compares the costs of both kinds of power generation systems including sensitivity analyses of the assumptions, and reaches some conclusions about what the future is likely to hold.

R. MICHAEL MURRAY, JR. *is a principal of McKinsey & Company in New York, and has served clients in the oil industry, utility industry, and electrical machinery manufacturing industry. He holds degrees in physics, mathematics, and business administration. Prior to joining McKinsey, Mr. Murray was a project engineer for the Naval Ship Research and Development Center.*

Problem Definition

Why "base load power generation?" Because these plants produce most of our electricity and hence have the biggest impact on our cost of electricity.

Utilities face a daily, weekly, and seasonal problem of matching, in the most economic fashion, the amount of electricity they produce to the amount their customers will demand. Within a day, demand falls off at night; within a week, demand peaks during work days; and within a season, demand depends on, among other things, climatic conditions of the area the utility serves. The utility's objective is to design its base load plant capacity to meet the minimum load that can normally be expected regardless of the variations in demand (usually 50 to 60 percent of peak load). The base load plants are typically large, new, coal- or oil-fired power plants, or nuclear plants or, in some locations, hydroelectric plants. They are the most expensive to build, but the least expensive to run, and overall produce the lowest cost electricity—if they are kept running at relatively high capacity levels.

When demand increases above the base load, the utilities bring on-line plants which produce increasingly higher cost electricity until, for the very peak in a day's demand, the utility usually has on-line equipment which produces its most expensive electricity. Table 1 shows typical capital costs, utilization, and electricity costs for base, intermediate, and peak load forms of power generation.

Most of the nation's electrical energy is produced from base load plants, and hence this chapter will focus on the economics of these plants.

Why "to the year 2000?" Enough justified controversy exists on the economics of power generation for the next twenty-five years; the economics of new power generation plants after that are truly unknown.

In 1975, the nation's utilities sold about 1,750 billion kilowatt-hours of electricity to their customers. This electricity was generated by a variety of power plants with a total capacity of about 500,000 megawatts. Kilowatt-hours of electricity sold per year have been increasing fairly steadily since World War II. Between 1955 and 1973, electric utility kilowatt-hour sales increased an average of 7.3 percent per year. Power generation capacity grew at a similar rate. In the years 1974 and 1975, the steady growth in kilowatt-hour sales was interrupted, apparently because of the higher costs of fuel, reduced growth in general economic activity, and

TABLE 1. COSTS OF POWER: BASE, INTERMEDIATE, AND PEAK LOAD GENERATING
         PLANTS

Kind of Plant	Typical Percent of Time Operating At Capacity	Typical Capital Cost of Capacity *	Typical Cost of Electricity Produced †
Base load — New oil, coal / Nuclear / Hydro	70	$350–480	14–22 mills
Intermediate load — Old coal, oil / Variable hydro / Combined cycle	40	210‡–240	25 ‡
Peak load — Gas turbine / Stored hydro / Diesel	10	100–120 §	40–50 §

* Dollars per kilowatt.
† Mills per kilowatt-hour.
‡ For new combined cycle (gas and steam turbine) plants.
§ Gas turbine plants.
Source: FEA Project Independence.

effective conservation measures. This interruption of growth rates has led to new and lower forecasts of electrical needs and, understandably, utilities have responded by delaying additions to their power generation capacity in accordance with these new forecasts. (For example, many utilities have delayed plants in the earlier stages of construction and/or deferred groundbreaking on new plants.)

The interruption of growth, new forecasts, plant delays, and new awareness of the need for energy conservation have led to serious discussions about how much electric power the nation will and should need over the next twenty-five years. It is not the purpose here to engage in such discussion; no matter what "reasonable" forecast is developed, the additions to the utilities' existing base of power generation capacity will have to be enormous.

As mentioned earlier, utility kilowatt-hour sales have grown at an average compounded rate of over 7 percent for some time. If electricity sales grow at 6 percent between 1976 and the end of the century, utility sales in the year 2000 will be about 7,500 billion kilowatt-hours—over four times last year's sales. If the growth rate is only 4 percent—a figure which most knowledgeable people regard as unrealistically low—utility sales of the year 2000 will be about 4,600 billion kilowatt-hours, over 2.5 times last year's sales.

Estimating the capacity additions needed to generate these increased amounts of electricity is rather complicated. The factors to be considered include estimates of new plant efficiencies (how much power new plants will deliver compared to stated capacity), effect of attempts to level demand peaks (whether consumption patterns can be changed to "level" utilities' load), and the need for reserve margins (essentially the utilities' safety factor). However, assumptions about these and other factors can be made; and if kilowatt-hour sales grow at 5 percent, midway between the above mentioned figures, the utilities will need about 1.5 million megawatts of capacity to deliver the electricity demand for the year 2000. Factoring in the need to replace some obsolete plants means that over two times today's power generation plant capacity would have to be added in the next twenty-five years to meet an electricity growth rate that many forecasters would agree is conservative.

Why is the choice essentially "between coal and nuclear power generation?" The choice exists because it can be assumed that national policy (if not simple economics) will prevent significant amounts of oil or natural-gas-fired power generation plants from being added to the system; because the amount of base load hydropower that could be added is limited by geographic and environmental considerations; and because the lead times required to develop, test, and build other nontraditional forms of power generation effectively preclude significant commercial application before the end of the century.

From the perspective of 1976, it is unlikely that the nation will rely on oil and natural gas for a significant percentage of its base load power generation capacity additions. As long as alternative sources are available (and they are in the form of coal and nuclear) the current 30 percent of capacity fired by oil and natural gas will steadily decrease as new plants are added to the system. Indeed, existing oil plants are already being converted to coal, and this conversion process can be expected to continue.

Hydropower generation capacity is currently about 67,000 megawatts—about 15 percent of the nation's total capacity. The 1.5 million megawatt system envisioned for the year 2000 includes 110,000 megawatts of hydro capacity, less than 10 percent of the system at that time. And part of the hydro capacity for the year 2000 includes hydro-pumped storage— a form of peak load capacity.

Among the more often mentioned nontraditional types of power generation are solar energy, geothermal energy, plants fired by liquids, or gas taken from coal or shale, breeder nuclear power generation and, finally, fusion power generation.

Solar energy could become an important source of energy to heat and cool homes and office buildings and thus provide an alternative to coal or nuclear power generation plants. However, most optimistic predictions for solar energy call for it to replace something quite a bit less than 5 percent of our traditionally sourced electrical energy by the year 2000.

Geothermal electrical power generation plants convert the heat within the earth to electricity. In 1974, FEA estimated there were 440 megawatts of capacity in operation—about 0.1 percent of total utility capacity. Under accelerated development, FEA has estimated that this could reach 7,000 to 15,000 megawatts of capacity by 1985, which, assuming a 5 percent per-year sales growth from 1975, would account for between 1 and 2 percent of total capacity in that year. Even with continued maximum development of this energy source, it is unlikely that it could account for more than 1 or 2 percent of capacity requirements by the year 2000.

Oil shale is potentially a very large source of energy. FEA reports that the reserves located in Colorado, Utah, and Wyoming are at least the equivalent of 1.8 billion barrels of oil. However, there are significant environmental problems associated with its recovery. FEA estimates that a one million-barrel-per-day shale oil industry would create 1.7 million tons per day of solid waste, would significantly alter the water table in the area of Colorado (where such an industry would be located), and would increase the salinity of the Colorado River. But even if these formidable environmental problems can be overcome, it would take a tremendous effort to develop a one million-barrel-per-day industry before the turn of the century, and even then it would be extremely unlikely that oil taken from shale could compete economically as a source of power generation energy. (To put the worth of a one million-barrel-per-day shale oil industry in perspective, FEA estimates that by 1985 the nation will be consuming about twenty million barrels of oil per day, plus or minus 10 percent depending on oil prices.)

The production of oil or gas from coal is a long way from commercial reality. Currently known technology would produce an extremely expensive (by today's standards) barrel of oil or its equivalent in gas. Prices estimated in 1976 dollars range up to twice the current international price for a barrel of oil. Estimates of lead times to develop "commercial scale" production vary, but an accelerated effort could yield a full-scale pilot plant with a 150,000-barrel-per-day capacity by the mid-1980s. The real problem appears to be that a fundamental change in technology may be needed to improve the comparative economics of the prod-

uct produced. This could put large-scale production well into the
1990s—not much help for electrical power generation through the rest
of this century.

Breeder reactor technology is also a long way from commercial application. Estimates vary, but the mid-1990s is about as soon as a significant amount of electricity is likely to be commercially generated
from this source in this country.

Nuclear fusion offers the potential for essentially unlimited energy.
It is the source of energy for the hydrogen bomb. But it is simply not
known whether nuclear fusion can produce energy in a controlled manner. ERDA has committed substantial funds for research into nuclear
fusion technology, as have similar institutions in other nations. By the
mid-1980s, the results of this effort may be known. However, even if
nuclear fusion should prove feasible by then, it will be well into the
next century before significant amounts of electrical energy from this
source will be available.

The foregoing is not at all intended to minimize the importance of
efforts to develop new sources of energy. On the contrary, it is critically
important that new sources be developed and become commercially
viable. FEA has estimated that without the development of major new
energy sources the nation is likely to be more dependent on imported
oil and gas in the year 2000 than it is today—even assuming substantial
and successful conservation measures, the rapid development of domestic
oil and gas reserves, and the expansion of coal and nuclear power generation to meet the nation's electrical needs.

However, the question facing the electrical utility industry is how it
should enlarge its power generation capacity to something around three
times its current size between now and the year 2000. The answer to that
question lies in the comparative economics of light-water nuclear reactor
and coal-fired power plants.

Background

In reality, power generation alternatives cannot be selected on
simple economics. The choice of coal or uranium is becoming increasingly influenced by what have been traditionally noneconomic factors
—although ultimately they express themselves in the economics of the
choice—a fact that is now becoming evident. For example:

  —Nuclear power plants have experienced tremendous cost increases—far in
    excess of what could be attributed to underestimation of the effects of infla-

tion or unknown plant construction problems. Many of these cost increases have been due to delays in the licensing process and changes in regulatory requirements. These cost increases or overruns can be thought of as an economic expression of the concern of many citizens about the safety of nuclear plants.

—Similar cost increases are occurring in other stages of the "nuclear cycle." As an example, industry estimates of the costs of reprocessing spent fuel and managing nuclear waste have about doubled in the recent past and are likely to increase again as industry and government gain understanding of what carrying out these functions really requires.

—Finally, nuclear plants have not operated at expected output levels. Plants have been shut down more often or forced to operate at lower levels of capacity than was expected. The reasons for this vary. Simple operating and start-up problems are among them; but some percentage of the unexpectedly poor performance has been safety related.

Coal-fired power generation plants have seen the same economic effect from "noneconomic" forces. Examples are:

—The productivity of coal miners had been improving steadily for some time: in 1960, the industry average for tons per man-day stood at 12.8; by 1969, it was 19.9. Productivity increases had come in both underground mining and surface mining. By 1973, the tons per man-day had fallen to 16.8, less for both types of mining. This reversal in productivity gains can be traced in large measure to the implementation of the Mine Safety Act passed in 1969. The fall-off in productivity has increased the cost of coal—an economic expression of a traditionally noneconomic fact.

—The availability of surface-mined coal and its cost have been adversely affected by environmental considerations and the requirement for restoration of exhausted surface mines.

—The cost of constructing coal-fired plants has increased substantially because of the EPA regulations limiting the amount of sulfur which coal plants may emit into the atmosphere. Estimates of this cost increase vary to as high as 25 percent of the total capital cost of the plant.

This is not to say that these noneconomic factors should not have become significant. Many of them, probably most of them, should have. It is rather to underscore a caveat that must be kept in mind for the rest of the chapter. The estimated costs associated with nuclear and coal power generation listed below are based on various public and private sources of information. Although these sources have learned from experience that they must take noneconomic factors into account, they admit that the economic impact of noneconomic factors can prove their estimates wrong.

Economics of Nuclear Power Generation

Before discussing in detail the "nuclear cycle" and the cost of electricity from nuclear power, it seems appropriate to set forth some basic facts about nuclear power today:

—By the end of 1975, fifty-two nuclear power plants were in operation with a total capacity of about 36,000 megawatts or about 7 percent of total utility generating capacity. In 1970, there were only 13 plants in operation; these plants had 5,000 megawatts of capacity or less than 2 percent of total capacity. Hence, in the scheme of things, nuclear power generation is a relatively new phenomenon and much is being learned about how to build and operate these plants efficiently.

—As mentioned above, nuclear power plants have not produced as much electricity as was originally expected. As a group, they have performed at 55 to 60 percent capacity factor. Capacity factor is the ratio of kilowatt-hours actually produced and the kilowatt-hours that theoretically could have been produced—that is, to achieve a 100 percent capacity factor, a plant would have to operate at "nameplate" rating (for example, 1,000 MWe) 24 hours per day all year long. One hundred percent capacity factor for a full year is impossible for any plant, but minimum expectations are for 70 percent capacity factors and operating objectives are for 80 percent and higher. Part of the cause for the historically low capacity factor experience is plant refueling. Refueling is done about once a year and in old plants can take over six weeks between shutdown and full operation or up to 15 percent of capacity. Another reason is "load following"—that is, base load capacity often is greater than maximum level demand so that base load plants have to be slowed down; this can account for another 10 percent of capacity. But 15 to 20 percent of capacity has been lost because of unscheduled shutdowns or other operating constraints.

—Despite these problems, nuclear plants have produced electricity which in general is cheaper than alternative sources of power. However, critics fairly argue that historical costs have been distorted by extremely low fuel costs and by failure to properly account for the cost of the full fuel cycle, among other things.

The difficulties encountered thus far in this new industry will be solved —the solutions are technically feasible. The economics will be discussed below. Many experts feel that the industry is poised on the edge of a learning curve; they point to the buildup of experience coming as new plants are built and come on-stream. There is no denying that the opportunity for learning is substantial and that the need matches the opportunity. There are about one hundred fifty nuclear plants on order.

Over one hundred of them will be completed within the next ten years, and these plants should provide a valuable basis for improving the performance of nuclear plants.

THE NUCLEAR FUEL CYCLE

Figure 1 is a simple schematic of the so-called nuclear fuel cycle. It shows, among other things, the uncertainty associated with the end of the fuel cycle. Table 2 is a brief explanation of the cycle: a description of each step; the amount of material needed by (or produced by) a typical 1,000 megawatt capacity (MWe) light-water reactor nuclear power plant; the current status of each step; the kinds of institutions involved in each step; and, finally, some of the expansion issues associated with each step. Much of the material for Table 2 was taken from an Atomic Industrial Forum, Inc. paper entitled "The Nuclear Fuel Cycle." The steps in the fuel cycle and the economic problems associated with each step are as follows:

Mining—This step may represent one of the biggest short-term limitations to a significant expansion of nuclear power generation. The problem is not whether there is enough uranium in the ground in the United States and in other countries to serve the needs of a growing number of nuclear power plants. In fact, the 3.5 million tons U_3O_8 ERDA estimated known U.S. reserves are enough to fuel the thirty-year life times of about a thousand 1,000 megawatt light-water reactors (assuming spent fuel recycling) or about six hundred 1,000 megawatt light-water reactors (assuming no fuel recycling). Rather, the problem is that for quite some time the price of uranium was too low to motivate miners to exploit known reserves and to explore for new reserves. This situation turned around dramatically in the past several years. The price of uranium was under ten dollars per pound of U_3O_8 for many years. In the early part of this decade, the price began to move rapidly, and prices in excess of thirty-five dollars per pound for delivery in 1980 are now common. These new prices are encouraging expanded production. Unfortunately, it takes up to eight years to explore, find, and begin mining of uranium.

The net result is that there is likely to be a shortfall of domestically produced uranium ore by about 1980. The extent and duration of the shortfall depend on how fast the currently ordered plants come on-stream and how quickly domestic uranium production can be expanded. In recognition of this likely shortfall, the government has ordered the current embargo on imported uranium to be gradually eased to allow the nuclear power generation industry to expand.

Fig. 1. Nuclear Fuel Cycle

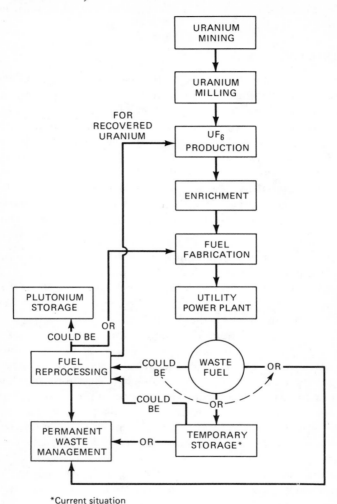

*Current situation

Milling—Ore is refined so it can be more economically shipped to the next stage of the cycle. Accordingly, milling is done very near the mine site. Milling capacity could also be short before 1980, but it has a shorter expansion lead time than mining and thus is unlikely to be a bottleneck.

UF_6—The next stage in the cycle is to turn "yellow cake," the output of the milling process, into a gas, UF_6. (A gaseous state is necessary for enrichment.) Few capacity problems are likely to exist in this step of the cycle.

Enrichment—For light-water reactors, it is necessary to increase the concentration of usable uranium isotope in UF_6 by a process of "enrichment." At present, the federal government performs the enrichment service for itself, domestic utilities, and foreign customers. Expansion of the enrichment capacity will be needed, possibly before 1985. Expansion is extremely expensive ($4 billion to $5 billion), but the capital costs of enrichment have to be viewed in light of the many power plants a typical unit of expansion can serve (about 125 plants). The basic problem is that expansion requires up to nine years and, hence, commitments to expand will have to be made soon. There are issues to be resolved concerning technology, ownership, and funding before expansion can begin.

Fuel Fabrication—After enrichment, UF_6 must be converted to a solid and then fabricated with other materials to make nuclear fuel. Fuel fabrication capacity is unlikely to be an early problem and expansion is relatively easy. The difficult question is what to expand: if fuel reprocessing goes ahead and if plutonium is recycled in power plants (see the discussion below), then fuel fabrication facilities will have to be built to produce fuel containing both uranium and plutonium.

Power Plant—In this step, the energy in the uranium is converted to electricity. The economics of this process will be discussed in the next section.

End of Cycle—As nuclear fuel is consumed in the process of producing electricity, it builds up fission products. These products effectively slow down the nuclear reaction in the power plant and therefore must be removed. Accordingly, about once a year between one-fourth and one-third of the fuel load is removed and fresh fuel loaded in the reactor. However, when the "spent" fuel elements are removed, they still contain usable uranium isotopes. They also contain plutonium, which has been "bred" in the reaction process. These uranium and plutonium fuel elements are valuable; the question is how valuable, in light of the cost of recovering these potential energy elements and putting them in the form of usable fuel. The economics of reprocessing are in some dispute. Most experts used to think that reprocessing would be a major economic advantage to nuclear plant owners, providing them with an alternate source of low cost fuel. The picture has now changed. Whether or not reprocessing can be shown to have some economic payoff depends on the assumptions used, but few expect the economic advantages to be as significant

TABLE 2. EXPLANATION OF THE NUCLEAR FUEL CYCLE

Step	Description	Amount Needed to Serve 1,000 MWe Reactor
Mining	Underground mining and surface mining of ore	263,000 tons per initial core 94,000 tons per reload
Milling	Mechanically and chemically refine ore to "yellow cake." Usually done near the mine	526 tons per initial core 187 tons per reload
UF_6 production	Convert "yellow cake" to gas for enrichment	420 metric tons uranium (MTU) per initial core 150 MTU per reload
Enrichment	Concentrate natural uranium content of U-235 at 0.7% to between 2% and 4% concentration Current technology being upgraded and new technologies being tested Modern plant with capacity of 9 million separative work units (SWU) requires about 2,500 MWe electric plant to run	9 million SWU plant serves about 125 power plants

Current Status	Institution Involved	Expansion Issues
9 million tons capacity 7 million tons production	Independent mining companies Large resource companies	U.S. known and estimated reserves are about 3.5 million tons of U_3O_8 (ore mined today has about 4 pounds U_3O_8 per ton of ore) at up to $30 per pound U_3O_8 It takes about 8 years to explore, find, and begin mining ore Major expansion effort needed—could require between $1.5 billion and $3+ billion in capital
18,000 tons capacity 12,000 tons production	Miners and chemical companies	Run out of capacity at a minimum by 1980, possibly within 2 or 3 years
16,000 MTU capacity (plus 9,000 MTU coming on-stream by 1977/1978	Chemical company Resource company	At minimum, no expansion needed by 1980 if planned expansion comes on stream By 1985, minimum need is for another 10,000 MTU capacity Accelerated program could require 10,000 MTU capacity by 1980 $30 million capital per 5,000 MTU block of capacity
Existing capacity plus planned upgrades could last until mid-1980s	Federal Government Private ownership being encouraged	A 9 million SWU plant costs between $4 billion and $5 billion and takes up to 9 years from design to start-up Depending on foreign commitments, capacity could be short before mid-1980 which would require early start at expansion

TABLE 2. EXPLANATION OF THE NUCLEAR FUEL CYCLE *(cont)*.

Step (cont.)	Description (cont.)	Amount Needed to Serve 1,000 MWe Reactor (cont.)
Fuel fabrication	Convert enriched UF_6 gas to solid and assemble in "burnable" fuel rod elements	90 MTU per initial core 25 MTU per reload
Utility power plant	Converts energy in uranium to electricity	Not applicable
Waste fuel	"Burned" up fuel bundles which if left in reactor can significantly reduce ability of reactor to sustain chain reaction needed to operate Has concentration of slightly less than 1% U-235 plus contains about 0.6% plutonium "bred" in the reactor	Produces about 25 MTU per year, total weight about 50 tons
Fuel reprocessing	Objective is to recover usable uranium and plutonium from waste	See waste fuel above
Waste management	Problem is high-level waste whether recycling proceeds or not. Problem is safe waste management essentially forever because of the level of radiation and the long life of the radioactive isotope	One year's output when reprocessed produces about 10,000 gallons of waste which when solidified and mixed with inert material take up about 70 cubic feet.

Current Status (cont.)	Institution Involved (cont.)	Expansion Issues (cont.)
Current capacity about 4,500 MTU	Nuclear steam system suppliers, oil company, others	Ultimate capacity and nature of capacity depend on whether plutonium will be recycled At minimum about 2,000 MTU more capacity needed by 1985; maximum up to 8,000 MTU new capacity About $140 million would meet minimum capacity requirements to 1985
52 plants with 36,000 MWe capacity About 150 plants on order with at least 100 becoming operational before 1985	Public utilities and federally owned utilities	To be discussed later in chapter
1,000 MTU backlog in on-site storage pools (temporary storage at power plant) More backlog stored at nonoperating reprocessing facilities	Public utilities and federally owned utilities	What to do with it and when: • Reprocess • Temporarily store and then reprocess it • Finally, permanently manage Expansion of temporary storage will be needed to handle waste fuel
No commercial operations now Planned capacity about 2,300 tons by 1980	Private companies: chemical and nuclear service	Will reprocessing be economic, i.e., will the value of usable fuel produced by recycling, reenriching, fabricating, etc., be greater than the cost incurred to do it—or should waste fuels be processed and sent directly to permanent waste management?
Federal Government will decide how to manage high-level wastes	Federal Government	Issue must be resolved

Source: Compiled from information in *The Nuclear Fuel Cycle,* June 1975 (Washington, D.C.: Atomic Industrial Forum, Inc.).

as was once thought. In any event the advantages must be assessed in the light of the problems of security and proliferation posed by "plutonium recycle" which are discussed in Chapters 1 and 5.

Waste Management—As discussed in Chapter 1, the ultimate disposition of high-level nuclear waste is an unresolved issue, except that the federal government will assume responsibility for them. The economics of waste storage will depend on the ultimate choice among several alternatives being considered, such as storage in canisters above ground cooled either by water or air, and permanent underground storage in geologically stable sites. For the present time, spent fuel is being temporarily stored in large pools of water located at power plant sites and other facilities provided for this purpose.

Summing up the nuclear fuel cycle, there appears to be two potential bottlenecks in the cycle for the next ten years:

—Mining, where in spite of current rapid expansion, there will probably be a shortfall necessitating imports of uranium at least until around 1980.
—Enrichment, where expansion will be needed by 1985 and where lead times require early resolution of ownership, technology, and funding issues if the capacity requirements are to be met.

Finally, the fuel cycle has to be ended either by reprocessing and permanent waste management or by no reprocessing and permanent waste management. There is less urgency in reaching an answer on how to close the cycle because there is available the alternative of temporary storage.

CONSTRUCTION AND OPERATION OF THE POWER PLANTS

In General—The methodology used for estimating the cost of electricity (from either nuclear energy or coal) is the same and involves two processes: estimating the cost to (1) build the power plant and (2) generate electricity from it.

Capital Cost—The cost is the total cost the utility has spent to build the plant, including the cost of borrowing money during the construction cycle. These costs are usually expressed in dollars per kilowatt of capacity.

When a utility decides to build a power plant, the management has estimates of hardware, construction material, labor, professional services costs, etc. The utility also has a construction schedule which specifies when certain pieces of the work will be completed, the payment schedule for services and equipment, and of course a scheduled date for beginning to generate electricity. Initial estimates of equipment, construction, and service costs are usually subject to escalation clauses to protect the suppliers from unexpected increases in inflation or costs incurred due to delays. The actual cost of escalation and interest in the ultimate capital cost of the plant is obviously time dependent. Hence the difficulty in estimating capital costs for power plants is twofold:

—The kinds of equipment and the construction and design effort that it will ultimately take to build the plant are not known at the outset of the project. Changes from the original estimate of what is required could come from evolution of equipment during the building cycle, unforeseen construction and design difficulties, and changes in regulatory requirements. In the case of nuclear power plants, where the technology has been rapidly evolving and where it takes ten years or more to build the plant, the uncertainty in capital costs is quite high (as is evidenced by the forecasting track record of the past ten years).

—The actual length of time it will take to build the plant is the second area of difficulty. Delays are common in any large construction endeavor; they have been especially significant in the building of power plants. Delays can and have come from many sources: equipment vendors not meeting schedules; lack of construction material or labor; unforeseen difficulties in building the plant; and finally, delays in construction due to regulatory changes and/or difficulties encountered in licensing.

Generating Costs—The cost of actually producing electricity has three basic components: the cost of the capital tied up in the plant itself, the operating and maintenance costs of the plant, and the cost of fuel. Generating costs are expressed in mills per kilowatt-hour of electricity produced. Hence, one major area of uncertainty is how much electricity the plant will produce, that is, how well it will run and at what capacity. The other areas of uncertainty are:

—Capital cost itself, that is, how much capital will be tied up in the plant once it begins to run.
—Cost of fuel, that is, what will happen to fuel prices over the life of the plant.

The operating and maintenance costs of the plant are fairly predictable and in any event are a small part of total generating costs.

For two categories, operating and maintenance costs and fuel costs, a technique called "levelizing" is used. Simply put, this technique attempts to even out the effects of inflation and price increases over some time period in the plant's life. (The calculations used here are levelized for the first ten years of plant operation.)

CAPITAL COST OF NUCLEAR PLANTS

Table 3 shows estimates for capital costs for nuclear power plants in two time periods, plants started in 1975 and plants scheduled for start-up in 1984. The estimates are based on an analysis of various data sources, public and private, and represent a consensus view of the data in these sources. They are expressed as point estimates, but actually represent a range of estimates.

TABLE 3.   CAPITAL COST ESTIMATES FOR NUCLEAR POWER PLANTS BY YEAR OF
           PLANT GOING INTO COMMERCIAL OPERATION

	Dollars/KWe 1975	Dollars/KWe 1984	Percent Change By Cost Category
Hardware costs	$110	$160	45
Construction costs	130	180	38
Professional service fees and other indirect costs	85	110	29
Interest during construction	70	230	228
Escalation	80	275	244
Total	$475	$955	101

Assumptions about 1984 costs:
   1. Length of building cycle (to commercial operation)—nine years
   2. Interest during construction—9 percent
   3. Escalation—7.5 percent (to 1980), 5 percent (1981-1984)

The 1975 plant capital cost estimate is based on a mix of actual and estimated plant costs. The statistical validity of actual plant costs would be weak because of the limited number in the sample.

The 1984 plant capital cost estimate is based on what various sources now think it will take to build a plant for operation by that date. These are the kinds of estimates currently being used by utilities to estimate capital costs on plants recently ordered or soon to be ordered.

Estimating costs of nuclear power plants beyond 1984 gets to be a game of extrapolation. For the purposes here, comparison of coal and nuclear power costs, extrapolation is not very instructive. Table 4 shows the percent of total estimated plant costs contributed by each cost category.

The message from these analyses is clear and obvious. Although the capital costs of nuclear power plants is expected to double by 1984, the main factors driving up the costs are interest and escalation. Interest and inflation rates are essentially uncontrollable, but time to construct is not. Efforts to standardize plant construction and design and to simplify (and strengthen) regulatory proceedings are under way now. The im-

TABLE 4. COST CATEGORIES AS PERCENT OF TOTAL NUCLEAR PLANT COSTS BY YEAR OF COMMERCIAL OPERATION

	1975	1984
Hardware costs	23%	17%
Construction costs	27	19
Indirect costs, etc.	18	11
Interest	15	24
Escalation	17	29
	100%	100%

portance of these efforts is critical. The actual cost reduction potential of reducing the construction cycle depends heavily on the assumptions used in how the plant will actually be built, that is, how, over time, equipment, material, and labor will be added to the plant. But cost reductions of up to 10 percent of total plant costs are not unreasonable expectations.

GENERATING COSTS OF NUCLEAR PLANTS

Table 5 shows estimated electricity generation costs for nuclear power plants beginning commercial operation in 1975 and 1984. Table 6 shows

TABLE 5. ESTIMATED GENERATING COSTS FOR NUCLEAR POWER PLANTS BY YEAR OF COMMERCIAL OPERATION

	Mills/KW-Hour	
	1975	1984
Capital charge	13.2	26.4
Operating and maintenance (O & M)	1.4	2.7
Fuel	7.4	11.4
	22.0	40.5

Assumptions:
1. Capacity factor—70 percent
2. Annual carrying charge—17 percent
3. O & M inflation—6 percent (to 1980), 5 percent (1981-1993)
4. Ten year levelized costs for O & M and fuel
5. Fuel prices in dollars per pound U_3O_8:

1975	1976	1980	1984	1990	1995
$25	$35	$46	$50	$65	$89

TABLE 6. MAKEUP OF FUEL COST ESTIMATES FOR 1984 LEVELIZED GENERATING COST

Cost Element	Percent
Refined ore	60
UF_6 production	2
Enrichment	25
Fuel fabrication	12
Reprocessing and waste management	13
Fuel credit	(12)
	100

how the various cost elements in the nuclear fuel cycle are estimated to contribute to the levelized fuel cost. Table 7 shows the sensitivity of generating costs to changes in assumptions. The sensitivity factor is critical.

Generating costs are very sensitive to the capital cost assumption—hence the importance of reducing the time required to build nuclear

TABLE 7. SENSITIVITY OF 1984 ESTIMATED NUCLEAR PLANT GENERATING COSTS TO CHANGES IN KEY COST ELEMENTS

Percent Change in Cost Element	Cost Element	Percent Change in Total Generating Costs
20	Capital cost	13
20	Capacity factor	13
50	Fuel cost	14
100	Refined ore	17
50	Enrichment	3
300	Reprocessing and waste management	7
Eliminate favorable credit	Fuel credit	3

Rationale for percent changes used:

Capital cost:	± 20% normal range for major projects
Capacity factor:	+ 20% yields 84% capacity factor
	− 20% yields 56% capacity factor
Fuel cost:	Overall uncertainty in fuel costs
Refined ore:	Uncertainty based on recent and forecast price movements
Enrichment:	Possible impact of private ownership of enrichment
Reprocessing:	Uncertainty of reprocessing and waste management costs
Fuel credit:	No reprocessing

power plants. They are also sensitive to capacity factors, that is, how the plant will run over a ten-year period. The assumption used in the calculation was 70 percent capacity factor. Recent experience has averaged closer to 60 percent. As the chart shows, a 20 percent change in this assumption (fourteen points of capacity factor) is worth 13 percent of generating costs—an obvious indicator of the importance of learning how to keep plants operating efficiently at high levels of output.

Generating costs for nuclear plants are not very sensitive to major swings in the cost of fuel. The overall cost of fuel would have to change by 50 percent to have the same impact as a 20 percent change in the capacity factor or the capital costs. And the costs of the elements which make up fuel costs would have to undergo dramatic change before they would have a significant impact on the generating costs of electricity from nuclear power plants. To illustrate this, consider the following:

—Table 5 shows the uranium prices, over time, used in calculation of levelized fuel costs. Forecasting prices of fuel has recently proved rather difficult. These prices are thought to be quite conservative—especially in light of the recent price history. However, they would have to double to have the same kind of impact on generating costs as a 20 percent change in capacity factor or capital costs.

—Enrichment is a capital intensive process—but the fact that one modern enrichment plant can serve about 125 large nuclear plants means that substantial changes in the price of enrichment have little impact on generating costs.

—The economics of the end of the fuel cycle are also unlikely to have major impact on generating costs. The assumption used makes the combination of reprocessing, waste management, and fuel credit a slight cost to the fuel cycle. But if there is no fuel credit, and reprocessing and waste management costs are three times the estimate used in calculation, the cost impact on generating costs is only 10 percent.

The message to be taken from this discussion is not that nuclear power generating costs will be 40.5 mills per kilowatt-hour in 1984; rather it is that capital costs of nuclear plants will about double for plants going into operation in 1984 when compared with plants having recently gone into operation; that generating costs will increase by less than the capital costs; that they are most sensitive to capital cost assumptions and operating efficiency; and that major changes in fuel costs have little effect on the costs of nuclear power.

One more point should be made before moving on to a discussion of coal generating economics. The estimates given in this section are for "national average" plants. However, nuclear generating costs should not vary significantly because of purely regional effects. The cost of nuclear fuel is essentially independent of the location of the power plant, as are hardware costs. Construction costs do vary—apparently dependent mainly on climatic conditions. As an example, using a national index of 100 for construction costs, the index would be 105 for the Northern States (New England, Middle Atlantic, East and West North Central, and Mountain) and 88 to 93 for the Southern States (South Atlantic, East South Central, West South Central, and Pacific). Hence, on the basis of the costs shown in Table 3 (see page 72), generating costs in the South Atlantic States would be slightly below the national average and generating costs in New England would be a little above the national average.

Economics of Coal Power Generation

BACKGROUND

Before discussing the coal cycle and the cost of generating electricity from coal-fired plants, some background may be useful:

—In 1974 coal-fired plants contributed 176,000 megawatts to the existing domestic generating capacity—about 38 percent of the total.
—Coal-fired power plants had traditionally produced about 70 percent of the nation's electricity, but in 1955 coal's share of electricity began to decline fairly rapidly to a low point in the early 1970s of 55 percent. (The apparent inconsistency is due to the fact that base load plants produce a disproportionately higher share of electricity than nonbase load plants.) Coal's share will now begin to increase again as coal starts to replace gas and oil in new additions of capacity.
—As has been the case with nuclear plants, the newer coal plants on average have not performed up to expectations. The capacity factors for large coal plants in the early part of this decade averaged slightly less than 60 percent, whereas earlier capacity factors for coal plants averaged above 80 percent. The reasons for this decline are varied:
    —A number of large new coal plants were brought on-stream during the early 1970s and new plants usually need several years to reach peak operating efficiency.
    —SO_2 removal devices were installed during this period, and may have contributed to lower capacity factors.
Nevertheless, the technology of coal-fired power plants is relatively mature, especially compared with nuclear technology. The recent problems will be overcome as utilities learn how to better manage the large plants with SO_2 removal facilities and as the plants themselves mature. Capacity factors above recent levels are reasonable objectives.

THE COAL CYCLE

Compared to the nuclear fuel cycle, the coal cycle is relatively simple.

Mining—Domestic minable reserves are enormous. FEA estimates that there are enough reserves which could be exploited at current price levels to last over 500 years. Clearly, supply is not a problem. However, there are some obstacles to the use of these reserves in coal-fired power plants.

—Although 60 percent of the nation's reserves have 1 percent or less sulfur content by weight, a smaller percentage of these reserves can meet the new EPA standards for SO_2 emissions established pursuant to the Clean Air Act Amendments of 1970.

—Much of the low-sulfur coal which could be used is relatively far from major consumption centers, especially on the East Coast, and transportation costs are a major factor in fuel costs.

—Much of the low-sulfur coal is in reserves accessible by surface mining. Surface mining faces fairly stringent environmental regulations which will impact on cost and availability.

Environmental issues aside, the coal mining industry faces a major management task merely to expand mining capacity to meet the currently forecast demand for coal. Coal production has expanded rather slowly in the recent past. Between the early 1950s and mid-1960s it increased only about 1 percent per year. Between the mid-1960s and early 1970s it has grown at slightly over 2 percent per year.

On the basis of the 5 percent kilowatt-hour growth, which was assumed earlier (see page 58), the coal industry would have to more than double the rate of production increases of the recent past if coal-fired plants are to account for as much as 50 percent of capacity additions over the next ten years. And the coal industry would have to continue to increase production at a rate of about 4 percent per year for the rest of the century to keep up with demand. If one assumes that coal-fired plants will account for 75 percent of capacity additions, coal production will have to expand by over three times the rate increases of the recent past through 1985 and by about 5 percent to the end of the century. To meet even the lower of these expansion requirements (that is, to account for 50 percent of capacity additions) will be a major undertaking. The capital cost associated with such an effort is estimated at between fifteen billion and twenty billion dollars over the next ten to fifteen years.

Transportation—Most of the nation's coal is transported by rail. Transporation costs are a major part of the purchase price of coal. As an example, in mid-1975 the average price of delivered coal paid by utilities was $0.81 per million BTUs of energy content. In New England, relatively far from coal mines, the price was $1.16 per million BTUs; in the Middle Atlantic region, the price was over $1.00 per million BTUs; and in the East Coast Central region the price was at the national average. However, in the Mountain States where utilities are located very close to low-cost sources of low-sulfur, surface-mined coal, the price was only $0.31 per million BTUs.

The railroads will also have to expand capacity dramatically over the next 10 to 15 years. Well over 100,000 new railroad cars will have to be added and the engine fleet be significantly expanded. Expansion could require capital in excess of three billion dollars over the next ten to fifteen years. Clearly this represents a major management and financial challenge to the nation's railroads.

Waste Disposal—There are three waste streams associated with burning coal in power plants: gaseous and particulate stack emissions, ash and slag from the combustion process, and sulfur removal. Stack emissions have been significantly curtailed. Ash and slag represent little problem and can be used as aggregate for construction materials or land fill. But removing sulfur from stack gases creates a disposal problem.

The most common process for removing sulfur from stack gases involves "scrubbing" the gases with lime or limestone solutions. The by-product of this process is a relatively insoluble calcium sulfate-sulfite mixture and magnesium sulfur compounds (the latter soluble in water). Large amounts of this mixture are produced by coal-fired power plants employing scrubbers to reduce SO_2 emissions. There are relatively few plants with scrubbers operating in the country today; hence the waste disposal problem is not severe and is being met by temporary measures. However, a future problem must be solved.

Summing up, the major problems are the environmental trade offs both at the mine and the power plant sites, and the difficulties in finding capital and management resources equal to the task facing the mining and railroad industries.

CAPITAL COST OF COAL-FIRED PLANTS

Table 8 shows the capital cost estimates for coal-fired plants in the two time periods used for nuclear plants, start-up in 1975 and in 1984. The sources of estimates and methodology are similar to those used to develop estimates for the nuclear plants.

The costs for 1984 plant start-up are intended to represent those that utility managers will be using in 1978 for new plant decisions. The size

TABLE 8.   CAPITAL COST ESTIMATES FOR COAL POWER PLANTS BY YEAR OF PLANT GOING INTO COMMERCIAL OPERATION

| | *Dollars/KWe* | | *Percent Change by Cost Category* |
	1975	*1984*	
Hardware costs	$ 65	$ 95	46
SO$_2$ removal	70	110	57
Construction costs	110	155	41
Professional services and other indirect costs	60	80	33
Interest during construction	55	175	218
Escalation	60	195	225
Total	$420	$810	92

Assumptions about 1984 costs:
1. Cycle to commercial operation—6 years
2. Interest during construction—9 percent
3. Escalation—7.5 percent (to 1980), 5 percent (1981-1984)

of the plants for which these costs apply is about 800 MWe capacity.

Table 9 shows the percentage of total estimated costs contributed by each cost category. As in the case of nuclear plants, capital costs are expected to increase substantially—almost doubling over the period. The main factors are again interest and escalation. However, it is unlikely that much time can be saved out of the six-year coal plant building cycle. The technology of constructing coal plants is relatively mature (at least when compared to nuclear plants), and any learning curve benefits from this point will be slow in coming.

The economic cost of environmental protection is also clearly shown in these tables: direct costs are 17 percent for the 1975 plant and 13 per-

TABLE 9.   COST CATEGORIES AS PERCENT OF TOTAL COAL PLANT COSTS BY YEAR OF COMMERCIAL OPERATION

	1975	*1984*
Hardware costs	16%	12%
SO$_2$ removal	17	13
Construction costs	26	19
Indirect costs, etc.	14	10
Interest	13	22
Escalation	14	24
	100%	100%

cent for the 1984 plant. The total cost of environmental protection is actually higher than these amounts because the direct costs must carry their share of escalation and interest costs.

GENERATING COSTS OF COAL PLANTS

Table 10 shows estimated generating costs for coal-fired power plants beginning commercial operation in 1975 and 1984. The methodology used here was similar to that used in developing generating cost estimates

TABLE 10. ESTIMATED GENERATING COSTS FOR COAL-FIRED POWER PLANTS BY YEAR OF COMMERCIAL OPERATION

	Mills/KW-Hour	
	1975	1984
Capital charge	11.6	22.5
Operating and maintenance (O & M)	2.8	4.5
Fuel	10.4	19.4
Fuel inventory	1.0	1.8
	25.8	48.2

Assumptions:
1. Capacity factor—70 percent
2. Annual carrying charge—17 percent
3. O & M inflation—6 percent (to 1980), 5 percent (1981-1993)
4. 10-year levelized costs for O & M, fuel, and fuel inventory
5. Fuel prices: $0.81/millon BTUs in 1975; 3 percent real price increase 1975 to 1984, no real price increase after 1984; 6 percent inflation 1975 to 1984, 5 percent inflation after 1984

for nuclear plants. Fuel inventory costs represent the charge for capital tied up in coal inventory. Table 11 shows the sensitivity of generating costs to changes in assumptions. Generating costs are less sensitive to capital costs and capacity factor and more sensitive to fuel costs. Moreover, SO_2 removal has a major impact on generating costs.

As an example—assuming that the same regional differences in coal prices mentioned earlier continue—in the Northeast where SO_2 removal is required and where coal costs are comparatively high, generating costs for coal plants are about 20 percent above the national average. In the Middle Atlantic and South Atlantic regions the costs are about 10 percent above the national average—again, assuming SO_2 removal costs as estimated. On the other hand, in certain parts of the Pacific Coast

TABLE 11. SENSITIVITY OF 1984 ESTIMATED COAL-FIRED PLANT GENERATING COSTS TO CHANGES IN KEY COST ELEMENTS

Percent Change in Cost Element	Cost Element	Percent Change in Total Generating Costs
20	Capital cost	9
20	Capacity factor	9
50	SO$_2$ removal plus share of interest and escalation	5
20	Fuel and fuel inventory costs	9
40	Fuel and fuel inventory costs	18

Rationale for percent changes used:

Capital cost:	$\pm\, 20\%$ normal range for major projects
Capacity factor:	$+\, 20\%$ yields 84% capacity factor
	$-\, 20\%$ yields 56% capacity factor
SO$_2$ removal:	Range of uncertainty in costs
Fuel and fuel inventory costs:	Range of costs for delivered coal paid in 1975

region, coal generating costs would be about 15 percent below the national average under the same set of assumptions. And, finally, for coal plants located near low-sulfur coal and with very low SO$_2$ removal costs, generating costs could be over 30 percent below the national average.

Concluding this section, the generating cost of coal will about double between 1975 and 1984. Reduction in capital cost will make some improvement in coal costs, but given the nature of technology involved significant improvements are unlikely. Increase in capacity factors, which should be possible, will improve coal generating costs somewhat. However, the major swing factors are whether SO$_2$ removal is required, how much it will cost, and what regional differences will exist in the delivered price of coal.

Comparison and Conclusions

Tables 12 and 13 present the comparisons of coal and nuclear generating costs. Table 12 shows the straight comparisons. Table 13 lists some sensitivity analyses by cost factor changes—that is, capital costs, capacity factor, nuclear fuel costs, SO$_2$ removal costs, and coal costs.

Capital Costs—If the base cost assumptions developed are reasonably accurate—at least in comparative terms—nuclear capital costs would have

TABLE 12.   COMPARISON OF GENERATING COSTS: NUCLEAR VERSUS COAL
"NATIONAL AVERAGE"

	Mills/KW-Hour	
	1975	1984
Total generating costs: nuclear	22.0	40.5
Total generating costs: coal	25.8	48.2
"Advantage" of nuclear over coal	17.3%	19.0%

to be 30 percent above the estimate for coal generating costs to come even. Actually, in any given plant this could happen. If plant construction—especially toward the end of the building cycle—were delayed a year or two, the total capital costs could easily be up by 20 percent. And if inflation rates are higher during the build cycle than those forecast, nuclear's "advantage" over coal will diminish. The opposite will happen if inflation is less than predicted.

Alternatively, if one assumes that there is more room for improving capital costs of nuclear plants than coal plants, nuclear's "advantage" widens. Such improvement could take place if the nuclear build cycle were reduced. There is direct evidence that this is possible. Large nuclear plants have been built in Europe in fewer than nine years, often in six to seven years.

The Capacity Factor—If nuclear and coal-fired plants continue to perform at roughly 60 percent capacity factors, then the nuclear advantage narrows. If coal plants improve to 80 percent and nuclear plants stay at 60 percent, then nuclear's advantage disappears. Alternatively, if both plants improve, then the advantage for nuclear increases.

Nuclear Fuel Costs—For nuclear to lose its advantage over coal, fuel costs would have to be almost 70 percent higher than estimates. But if costs for both kinds of fuel increase faster than estimated, nuclear's advantage widens. This is the kind of situation which could prevail under energy shortages or if inflation, after the plants are built, is higher than estimated.

SO_2 Removal Costs—SO_2 removal costs can easily change nuclear's advantage. Removing the equipment and associated indirect costs would reduce nuclear's advantage to 9 percent.

Regional Coal Costs—If there is no need for SO_2 removal systems and

TABLE 13. SENSITIVITY OF 1984 NUCLEAR "ADVANTAGE" OVER COAL TO CHANGES IN KEY COST FACTORS

	Change in Cost Factor	1984 "Advantage" Nuclear over Coal
Capital costs	Base case ("national average")	19%
	29 percent increase in nuclear capital cost; 0 percent increase in coal capital cost	0
	20 percent increase in capital cost of both	15
	20 percent reduction in capital cost of both	24
	20 percent reduction in nuclear capital costs; 10 percent reduction in coal capital costs	31
Capacity factors	56 percent for both	13
	60 percent capacity factor for nuclear; 80 percent capacity factor for coal	0
	84 percent capacity factor for both	25
Nuclear fuel	67 percent increase in nuclear fuel cost; 0 percent increase in coal cost	0
	20 percent increase in nuclear fuel; no change in coal costs	11
	20 percent increase in both fuel costs	22
SO_2 removal costs	50 percent reduction in SO_2 removal; no change for nuclear	15
	Zero cost for SO_2 removal; 10 percent less escalation and interest costs; no change for nuclear	9
Regional coal costs	Zero cost for SO_2 removal; 10 percent less escalation and interest costs; 17 percent lower coal costs; no change for nuclear	0
	Zero cost for SO_2 removal; 10 percent less escalation and interest costs; 25 percent lower coal costs; no change for nuclear	(5)
	25 percent lower coal costs; no change for nuclear	6
	20 percent higher coal costs; no change for nuclear	29

if low-cost coal is available, nuclear's advantage can disappear—or become negative. Even if SO_2 removal systems are required, nuclear's advantage becomes insignificant when coal costs are 25 to 30 percent below the estimated national average.

SUMMING UP SENSITIVITY ANALYSES

The sensitivity analyses set forth above strongly indicate what the economic strategic objective for the nuclear power generation industry has to be: it must reduce the capital cost of effective plant capacity, that is, increase capacity factors while reducing or holding constant plant capital costs. To put this objective into perspective: if the performance of coal-fired plants improves so that they reach 70 percent capacity factors and all other coal costs are as estimated, and if nuclear plants cost only 10 percent more than estimated, with nuclear plant capacity factors staying at 60 percent while all other nuclear costs remain as estimated, then nuclear's advantage essentially disappears in the "national average" case.

Reducing the capital cost of effective plant capacity should be possible. It will involve:

—Trading off increases in plant capital cost for "assured" increases in capacity factors.
—Standardizing plant design and construction to reduce both the direct capital cost of plants and the length of time required to build them.
—Reducing the time spent in licensing nuclear plants.

Steps are being taken in this direction: the NRC has limited the size of plants so that manufacturers can focus their efforts on improving performance without worrying about size increases; standard plant licensing procedures are being implemented; and standard plants are being built—but it is too early to see the actual effect of standardization.

The economic objectives of the coal-fired generation industry are also relatively clear. Plant performance must be improved, SO_2 removal costs controlled and system effectiveness increased, and coal mining and transportation costs controlled.

One point on the economics of the end of the nuclear cycle should be made. The above analyses show that at least for the next fifteen years or so the economics of nuclear power generated electricity are not particularly sensitive to how the nuclear cycle is ended. Beyond the next fifteen to twenty years the price of nuclear fuel is likely to be significantly affected by whether or not a reprocessing industry exists to supplement the amount of uranium available from mining. But in the meantime, in the overall scheme of things, this is economically less important than learning how to reduce the capital cost of effective plant capacity. This might mean that the nation ought not to consider full-scale reprocessing during the next several years until federally funded and controlled large-scale pilot projects show how this can be done economically and safely.

Admittedly, delaying full-scale reprocessing will lead to higher costs; temporary storage facilities will have to be expanded, uranium prices could increase faster than expected, and a larger ultimate reprocessing capacity might be needed because of the build-up of waste fuel inventories in temporary storage. However, the economic cost of delaying full-scale reprocessing has to be weighed against the risks associated with proceeding too fast—and these risks involve safety, technology, economics, and politics; that is, the total future of nuclear power generation could be seriously jeopardized by even the most minor incident involving spent fuel reprocessing and permanent waste management.

CONCLUSION

By the year 2000, the nation's power generation capacity will have to be at least three times its current size. Meaningful additions to capacity during this time period can only come from light-water nuclear power plants or coal-fired power plants, assuming oil and natural gas will not be used.

The current assessment is that electricity from nuclear plants is, on the average, less expensive than electricity from coal plants. The difference is primarily a function of the availability to the utility of low cost, low sulfur coal. Therefore in some parts of the country coal-fired plants will produce the least expensive electricity and in other parts the two forms of power generation will be at parity. However, for much of the nation, nuclear power generation should be the most economic form of electricity production.

Delivering the nuclear economic advantage will require a concerted effort to reduce the capital cost of effective plant capacity. In addition to delivering this economic advantage, there are steps in the nuclear cycle where critical issues need to be resolved: on the capacity side, they involve mining and enrichment; on the safety and technology side, the issues are how to end the cycle. The resolution of the former set of issues is relatively straightforward—but there is a matter of timing if major bottlenecks are to be avoided. The resolution of the latter set of issues will be more difficult—but time is available.

The major objective, reducing the capital cost of effective plant capacity, should be achievable. There are now on order about 150 nuclear power plants, of which over 100 will be completed within the next ten years. The experience thus gained will provide invaluable (and heretofore unavailable) data which can be used to improve nuclear plant construction and operating performance.

Fritz F. Heimann

3

How Can We Get
the Nuclear Job Done?

Introduction

The development of nuclear energy requires unusual institutional arrangements. Both government and private industry must operate in ways which differ from the conventional conceptions of their roles. There are many reasons why this is so: the issue of public acceptance of nuclear power, including concerns about accidents and environmental effects; the enormous amounts of capital and long time cycles required to build nuclear power plants and other facilities; the historical evolution of the nuclear power program from the nuclear weapons program; the recurring concern about diversion of nuclear materials; and the relation between U.S. and foreign nuclear power programs and the issue of proliferation.

At the beginning of the U.S. nuclear program, there was extraordinary creativity in fashioning novel arrangements to meet the demands of nuclear development. The Manhattan District Project, the Congressional Joint Committee on Atomic Energy, and Admiral Hyman Rickover are among the more unusual institutions which were invented at that time. All three represented triumphs of pragmatism over ideology, of substance over form. Similarly, in the early days of the civilian power program innovative arrangements were developed, including the AEC's power

FRITZ F. HEIMANN *is Counsel for General Electric's Power Generation Group. The views expressed are his own and should not be attributed to the General Electric Company. In 1972 Mr. Heimann edited The American Assembly's book* The Future of Foundations.

reactor development program and the Price-Anderson "no fault" insurance system.

In the past, a spirit of optimism and enthusiasm produced creative solutions to difficult institutional problems, even though the objective need for developing nuclear power was then far from compelling. During the 1950s and 1960s domestic fossil fuel supplies were abundant, and vast new low-cost petroleum supplies were being opened up in the Middle East. To show that nuclear power could become competitive with fossil fuels required very optimistic projections of cost improvements, with only a limited technology base to rely on. Similarly, the initial assumptions regarding the safety of nuclear plants were based on much more limited data than are available at this time when we have a track record involving three hundred reactor-years of power plant operation, as well as two thousand reactor-years of naval experience.

The desirability of developing nuclear power was clearly enhanced by the energy crisis and the quadrupling of petroleum prices. Yet, we now find ourselves boxed in by disputes we seem unable to resolve, by conflicting priorities we cannot reconcile. Over half the nuclear power plants which were on order in the spring of 1974 have been deferred or cancelled. Regulatory disputes over siting, reactor safety, and environmental protection have resulted in a ten-year time cycle for the construction of nuclear power plants, compared with five or six years in the early 1960s. This, in turn, has sharply increased construction costs, with money costs—interest on capital tied up during construction—exceeding the hardware cost of the reactor and the turbine generator. The uranium market has been disrupted by rapid price increases. After a decade of debate over whether government or industry should build the next uranium enrichment plant, neither has done so.

The "back-end" of the fuel cycle is in even worse disarray. No spent fuel reprocessing plant is in operation in the U.S., and those under construction are unlikely to start up in the foreseeable future. New projects will not be initiated until regulatory issues crucial to the design, operation and economics of reprocessing plants are resolved. The plutonium issue is enmeshed in the debate over safeguards against diversion and the proliferation controversy. While the disposal of radioactive wastes has long been recognized as a key issue affecting the public acceptance of the nuclear option, basic decisions regarding the form in which waste should be stored and the location of storage facilities have not yet been made.

What makes the present disarray so deplorable is that it comes at a time when most experts have concluded that expansion of nuclear power, together with increased coal production, represent the only options which can make significant contributions to U.S. energy needs in the next twenty-five years. The development of advanced power sources, such as fusion, solar energy, and geothermal power will not result in commercially significant additions to the U.S. power supply in this century. Expansion of coal production involves serious environmental impacts and carries with it huge investment requirements. While increased conservation is essential, it will have only modest effects on the long-term growth of energy demand. Thus, in any objective comparison of the alternatives, the expansion of nuclear power as part of a balanced U.S. energy program has substantial advantages from both the economic and the environmental standpoint. Every other major industrial nation has elected the nuclear option.

Unlike earlier periods of optimistic expectations, there is now an extensive base of experience from which future growth can be projected. In the U.S. about sixty nuclear power plants were in commercial operation in the spring of 1976, with an electrical generating capacity of approximately 40 million kilowatts. This represents roughly 8 percent of our national generation capacity. According to a recent survey by the Atomic Industrial Forum, the existing nuclear power plants produced electricity at an average total cost [1] 40 percent less than the comparable cost of fossil fuel plants. Furthermore, each one million kilowatts of nuclear capacity can save ten million barrels of oil each year. There are currently 140 reactors under construction or with construction permits pending. These will add over 150 million kilowatts to the nation's electrical capacity. The engineering and manufacturing capability to proceed with the construction of these nuclear plants is in place, and their potential contribution to U.S. energy independence is clear. The real issue is whether we can make the political decisions necessary to resolve the uncertainties in which the U.S. nuclear program is now mired.

Because of the economic recession, we have had two years without significant increase in demand for electric power. This has deferred the pressure to make difficult political decisions. At a time when the division of political authority in Washington makes it hard to resolve any controversial issues, it is understandable that the nuclear debate has continued without resolution. However economic recovery is bringing with it a resumption in energy growth. Thus, the time for postponement and procrastination is coming to an end. Either we will make the decisions

---

[1] This total includes construction cost, fuel cost, and operating and maintenance costs.

necessary to get the nuclear job done or the opportunity to proceed may be lost by default.

It is obvious that the controversy over nuclear power reflects some very fundamental issues in American society. The lack of trust in large institutions, including both government agencies and private corporations, is a major factor underlying public concern about nuclear safety. In addition, attacks on traditional values, such as the importance of economic growth and rising standards of living, are among the real objectives of some opponents of nuclear power. How long-lasting or widespread these concerns are is hard to predict.

A somewhat more tangible issue involves the roles to be played by private industry and by the government in getting the nuclear job done. This is hardly a new issue. Although it has been debated many times, the discussion has rarely transcended the level of clichés. On one side it has been argued that nuclear power is a crucial national resource in which billions of dollars of government money have been invested, and that "selfish private interests" should not be permitted to reap enormous profits. The other side has argued that nuclear power should be turned over to private enterprise as quickly and as completely as possible, that only through private competition can the nuclear job be done in an economic and efficient manner. By this time it should be clear that neither of these simplistic views provides a practical guide to the actions which must be taken. Instead of relying on ideological formulas, it behooves us to examine realistically what private industry can be expected to do, and where government action will be required.

Our analysis is divided into two sections: *How We Got Where We Are* presents an historical overview of the U.S. nuclear program, focusing on the roles of government and industry. *Where Do We Go From Here?* discusses the actions required to make the nuclear option viable.

How We Got Where We Are

The history of the nuclear power business resembles a geological succession of warm periods when ambitious plans grew in an atmosphere of sunny optimism, followed by ice ages during which these plans shrivelled and many of the planners froze or returned to warmer climates. Before studying these geologic successions, we must look at the prior period of government monopoly from which the nuclear power industry emerged.

THE PERIOD OF GOVERNMENT MONOPOLY

Nuclear energy is unique in American economic history in having been initiated as a government monopoly. From its inception in the early days of World War II, the Manhattan District Project was conducted in total secrecy under the direction of the War Department. Working through leading universities notably Berkeley, Chicago, and Columbia, it organized the nation's nuclear scientists to conduct the fundamental research necessary to separate U-235, produce plutonium and build the initial bombs. The major engineering and production jobs, including the construction and operation of the uranium enrichment facilities at Oak Ridge, Tennessee, and the plutonium facilities at Hanford, Washington, were conducted under contracts with industrial companies, including DuPont, Eastman Kodak, and Union Carbide.

Among the legacies of the Manhattan District Project which are still with us, are the National Laboratories at Los Alamos and Argonne and the production complexes at Oak Ridge and Hanford. Even though the Manhattan District Project employed tens of thousands of people and spent $2 billion, its existence did not become a matter of public knowledge until after the bombs were dropped on Hiroshima and Nagasaki in the summer of 1945. Three decades later, public attitudes about nuclear energy are still influenced by the nature of its initial product and the way it was developed.

The first open political debate about the atom developed over the issue of civilian vs. military control and led to the passage of the Atomic Energy Act in 1946. In the light of subsequent history, it is ironical that the organization of the civilian-controlled Atomic Energy Commission was regarded as a great liberal victory. The 1946 Act established a formal statutory framework for maintaining the government monopoly. All nuclear materials and all facilities for the production and use of nuclear materials remained under government ownership. Strict security classification was continued and new inventions were "born secret." Studies of nuclear power or other "civilian applications" could be conducted only under AEC contracts. The 1946 Act also established the oversight role of the Joint Committee on Atomic Energy which it has exercised ever since.

Under the leadership of its first chairman, David Lilienthal, who had previously run the Tennessee Valley Authority, the Atomic Energy Commission continued the Manhattan District's contracting approach. Lilienthal preferred to rely on contractors rather than develop large "in house" government staff. Additional contractors, including both General Electric and Westinghouse, were brought into the program and several more university-related national laboratories were established. This approach was continued by AEC throughout its life and has been inherited by its successor, the Energy Research and Development Administration (ERDA).

AEC's activities during its early years were largely focused on military applications, including the production and testing of atomic bombs. The development of propulsion reactors for submarines became a major activity in the late 1940s. This program was directed from the beginning by Admiral (then Captain) Rickover. Operating with powerful support from the Joint Committee, Rickover established a largely autonomous program, nominally housed within the AEC and the Navy Department. Rickover also used the contractor approach. The Bettis Laboratory, operated by Westinghouse, began the development of the pressurized water reactor for Rickover's naval reactors branch. The Knolls Laboratory, operated by General Electric, first developed a sodium-cooled reactor and then switched to pressurized water technology.

OPENING THE DOOR TO PRIVATE PARTICIPATION

In the early 1950s the AEC sponsored a program of nuclear power studies under which several industrial groups were permitted to investigate various reactor types for electric power generation purposes. The study groups included utilities, architect-engineers and equipment manufacturers. Among the many reactor types studied were pressurized water and boiling water, the sodium-cooled breeder and gas-cooled reactors. During this period the AEC conducted a program of building small experimental reactors at the national laboratories.

In 1953, the Joint Committee on Atomic Energy began to debate a change in the law which would permit increased private participation. This led to the adoption in the following year of the Atomic Energy Act of 1954. The new law for the first time permitted private industry to build and operate nuclear plants on their own initiative, and not just as government contractors. The act established a regulatory framework which required that privately owned reactors be licensed by the AEC. AEC's licensing responsibility was primarily directed to the protection of public health and safety. It had no responsibility for economic regulation, except for encouraging competition. Nuclear materials remained under the government monopoly, but AEC was authorized to make enriched material available to private licensees. One of the stated purposes of the 1954 Act was to encourage widespread participation in the development and utilization of atomic energy for peaceful purposes.

The AEC organized a power reactor demonstration program under which it assisted manufacturers and utilities to enter the nuclear field. The forms of assistance varied from research assistance and waivers of lease charges on nuclear materials to support of construction costs. The Shippingport plant, the first pressurized water reactor designed for electric power generation, was built by Westinghouse under a government cost reimbursement contract. This project started before the 1954 Act, and was carried on under the wing of Rickover's naval reactor program. The AEC's power reactor program attracted a large number of participants including many companies which had not previously competed in the electrical equipment market. AEC also encouraged private development by declassifying large amounts of power reactor technology before the Geneva conference on peaceful uses of atomic energy in 1955.

An initial obstacle to increasing private participation was concern about liability from a nuclear accident. This obstacle was removed by the passage of the Price-Anderson Act in 1957. It enacted an innovative system, which involved major changes in the traditional approach of the insurance industry, the utilities, and the supplies industry.[2] It is worth noting that Price-Anderson was enacted during a period when the AEC was part of the Eisenhower Administration, while the Congress was under Democratic leadership.

The mood of the mid-1950s was one of widespread optimism and enthusiasm. Not only the AEC and the new industrial participants, but also the press and the public were enthusiastic about the prospects for nuclear power. It was commonly assumed that, after the successful operation of demonstration plants built with government assistance, private industry would stand on its own feet and build nuclear power plants which would compete economically with fossil fuel plants. In fact, General Electric built the first boiling water reactor (BWR) plant, the Dresden Station, for Commonwealth Edison without applying for AEC help. General Electric management proceeded in the confident expectation that it could develop the BWR technology without government entanglements and have a commercially competitive product by the 1960s.[3]

---

[2] For a discussion of Price-Anderson, see Chapter 4.

[3] The Dresden project was assisted by a $15 million contribution from a group of interested utilities.

THE FIRST ICE AGE

It soon became clear to the industry that the expectations of rapid commercialization of nuclear power had been overoptimistic. While the initial water reactor plants, including Shippingport, Dresden, and Connecticut Yankee, operated well, there were very few follow-on orders and it became apparent that it would take more than one round of demonstration plants to make nuclear power competitive with coal- and oil-fired plants. The late 1950s and early 1960s were a period of letdown and discouragement, during which a number of the early entrants dropped out of the nuclear business.

Government policy both at the Joint Committee and the AEC level remained one of strong encouragement to the nuclear industry. However, there was increasing concern about the competitive impact of AEC assistance, as well as reluctance to become involved in "bail outs" of unsuccessful private projects. AEC's assistance in the later reactor demonstration projects went to carefully defined research and development tasks, and not to plant construction costs.[4]

THE TURNKEY ORDERS

The first ice age came to an end when General Electric and Westinghouse both launched a series of so-called "turnkey projects" in the 1963-1965 period.[5] The turnkey approach had two objectives. First, it was designed to overcome the reluctance of the utilities to assume the uncertain risks of building nuclear plants. By taking turnkey responsibility under fixed price controls, the manufacturer assumed most of the risks which troubled the utilities. In addition, the turnkey approach enabled the reactor manufacturer to undertake the overall systems integration of the entire plant, including the work of the architect-engineer. The plants offered were in the 500,000 to 800,000 kilowatt range, compared with the 100,000 to 200,000 kilowatt size sold in the 1950s. By going to the larger sizes it became possible to project energy cost levels which appeared competitive with those of fossil fuel plants. The utilities ordered over a dozen turnkey plants.

The turnkey orders produced a second period of optimism in the

4 The "Modified" Third Round Demonstration program of 1962 provided for AEC support of engineering costs of "first of a kind" features.
5 Under a "turnkey" contract the manufacturer took responsibility for all aspects of the design, procurement and, construction at an entire nuclear power plant. In theory, the utility merely had to "turn the key" after the plant was completed.

nuclear business. Many speeches and articles announced that the age of competitive nuclear power had indeed arrived. As one symbol of the times, Congress enacted legislation in 1964 providing for private owner-ship of enriched material, formally ending the government monopoly on the ownership of nuclear fuels. Also during this "warm period" several private ventures into the nuclear fuel reprocessing field were launched. The period of optimism lasted only three years.

THE SECOND ICE AGE

By 1966 it became apparent that the manufacturers had taken much larger risks than they had anticipated. The full magnitude of the disaster did not become clear until the early 1970s when construction of the turnkey plants was finally finished.

During the period that the turnkey plants were being designed and built, three developments completely changed the outlook which pre-vailed when the contracts were taken. First, the Vietnam war effort resulted in rapid inflation, shortages of construction labor, and delivery delays for important materials and components. This caused large cost overruns and schedule delays. Second, the newly emerging environmental movement led to a host of new regulatory requirements at the federal, state, and local level. This caused additional delays, as well as large unanticipated expenditures.

Third, the AEC licensing process was radically transformed. During the first decade after the passage of the 1954 Act, reactor licensing gen-erally involved a relatively informal technical review process. Engineers from the AEC's regulatory staff would review the design of the proposed plant with the technical personnel of the reactor manufacturer and the utility operator. Requirements for additional studies or design changes emerged out of these technical discussions. The lawyers' role was usually limited to drafting the final licensing documents at the end of the technical review process, and conducting public hearings which were usually brief, uncontested formalities.

In the later 1960s environmental groups began to intervene in the AEC licensing process, and the number of interventions multiplied rapidly after the National Environmental Policy Act broadened the scope of AEC's responsibility to include environmental effects unrelated to radiation. The process became highly litigious. The lawyers for the inter-venors employed all of the procedural tactics customary to bitterly contested legal disputes. Public hearings dragged on for many many months, and in some cases for years. Rulings which went against the intervenors were appealed to the commission and to the courts. Initially neither the AEC regulatory staff nor the utilities were prepared to deal with this challenge, and for a time the intervenors almost brought the regulatory process to a standstill.[6] For the nuclear industry this transformation of the regulatory process meant long delays and substantial cost increases.

The combined effect of all three of these changes was to produce enormous cost overruns on the turnkey projects, resulting in losses to the manufacturers totalling hundreds of millions of dollars. The extra costs to the utilities, from the lengthy delays in the completion of the plants and from other costs which they bore, also totalled many hundreds of millions of dollars. In addition, disputes over the responsibility for cost increases and delays embittered relationships between manufacturers and utilities.

From a technical standpoint, the turnkey plants have been a great success. After overcoming the initial "teething pains," which normally accompany substantial technological advances, the turnkey plants have operated at availability levels which compare favorably with those of large fossil fuel plants. The nuclear plants of the turnkey era generally produce the lowest cost electricity on their utilities systems.

The losses incurred on the turnkey jobs forced the manufacturers to retreat to a more modest scope of work under which their responsibility was generally limited to the reactor hardware and nuclear fuel. While this reduced the manufacturer's financial exposure, it complicated the job of designing and building nuclear plants. Overall integration of the total plant system was made more difficult with troublesome consequences in terms of obstructing standardization and complicating the licensing process.

General Electric and Westinghouse stopped offering turnkey plants in 1966. In the late 1960s there followed another period of low order volume.

THE GREAT ORDER SURGE: 1970-74

In the early 1970s the manufacturers began offering reactors in the 1 to 1.2 million kilowatt size range, with very favorable plant and fuel cost projections. During the period between 1970 and the middle of 1974 an enormous wave of orders came in, totalling over one hundred

---

[6] It took the AEC over a year after the Calvert Cliffs decision to prepare environmental impact statements, dealing with effects other than radiation for the then pending backlog of license applications.

reactors. This meant that nuclear power had won more than half of all the orders for generating equipment placed by the utilities during this period. Once again the nuclear industry was in a state of great optimism, and it seemed, with solid justification. In addition to Westinghouse and General Electric, Combustion Engineering, Babcock & Wilcox, and General Atomic received substantial reactor orders.

During this period U.S. light-water reactor technology also became very solidly established in the world nuclear market. Westinghouse and General Electric, together with the foreign manufacturers licensed to produce their light-water reactors, obtained a large volume of orders in Western Europe and Japan. The light-water reactor became the world standard, outstripping the British gas-cooled reactors and the Canadian heavy-water reactors by a wide margin.

THE THIRD ICE AGE

The boom ended in the second half of 1974. The flow of orders suddenly reversed and the utilities began deferring and even cancelling nuclear orders. The combined effects of inflation and then recession placed the utility industry in a severe financial squeeze. By the summer of 1974 utility interest costs had reached an all time high, and efforts to obtain rate increases were encountering bitter opposition and long delays. Consequently it became increasingly uncertain how the utilities could raise the money required for all the new equipment they had ordered. Failure to obtain adequate rate increases made internal generation of funds difficult, and high interest costs made outside financing unattractive. At the same time, the recession put an end to the period of strong growth in electrical load, which had continued since the mid-1960s. For the years 1974 and 1975, electrical load, on an overall national basis, remained below the 1973 level. The combination of financial squeeze plus uncertainty about future generation requirements, resulted in the wave of deferments and cancellations which affected over half the nuclear plants on order in 1974. Thus, for the third time in its short history the nuclear industry cycled from euphoria to depression.

Each cycle has been more extreme than the preceding one. In the spring of 1976 it was far from clear when there would be another upturn. Electrical load growth is likely to resume as the economy pulls out of the recession. However, this does not assure a prompt resumption of utility orders. Substantial amounts of previously ordered generating capacity have gone on the line during the period when electrical load was not growing. As a result, many utilities have larger reserve margins than ever before in their history. They can afford to wait several years, and make sure there will be sustained load growth, before ordering new plants.

Even then it is uncertain how large a percentage of their orders will go nuclear. The political opposition to nuclear power has continued to mount. Thus, a utility which selects a nuclear plant can anticipate bitter opposition. Even when a utility can feel confident that a nuclear plant will ultimately be approved, it cannot be sure how long the dispute will last and when the plant will go into operation. The arena of controversy has expanded from the nuclear licensing process to state utility commissions, state legislatures, and initiatives and referendums. Because utilities are regulated by politically-appointed utility commissions, their ability to proceed with controversial projects is limited.

The obvious question must be raised: how long can the nuclear manufacturers continue? General Atomic dropped out of the nuclear power plant market in the fall of 1975. Even with the backing of two of the world's largest oil companies, Gulf and Shell, General Atomic's future prospects had become too precarious to justify continuing its effort to develop a high temperature gas cooled reactor.[7] The four remaining manufacturers of water reactors can still obtain some comfort from the large order backlog obtained in the 1970-1974 period. However, it has become increasingly clear that a large backlog provides no guarantee of a viable future. Orders can be deferred or cancelled. Moreover, orders for distant delivery dates may turn out to be one-way options under which the manufacturer may be required to build a reactor of an out-moded design, at a price which may be unprofitable in the light of inflation and reduced business volume. Even though the manufacturers operate in a less confining environment than public utilities, there is a limit to how long they can swim against the tide.

Where Do We Go from Here?

In considering the future of the nuclear option it is essential to remember that for the United States nuclear energy is indeed an option. Unlike Japan or Western Europe, the United States has other alternatives. We have enormous coal reserves, as well as substantial domestic petroleum supplies. There is no law of nature or of economics which dictates that the United States must use nuclear power. It follows, however, that failure to develop nuclear power in the U.S. will not prevent other

---

[7] General Atomic paid $200 million to terminate two remaining contracts after most of its other orders had been terminated by the utilities.

nations from proceeding. For most of the industrialized world, there is no practical alternative to building more nuclear plants. Thus, the issue is not "should we bequeath a nuclear world to our children?" but only "should the U.S. avail itself of the benefits and the burdens of nuclear energy?" This section will consider the steps which should be taken if the U.S. decides to pursue the nuclear option.

THE PUBLIC ACCEPTANCE PROBLEM

Fundamental to the viability of the nuclear option is the resolution of public concerns about the safety of nuclear power. Like other controversial issues in our society, the acceptability of nuclear power must be settled by the political process. Thus there is no alternative to opening public debate of the issues raised by the opponents of nuclear power. It is admittedly difficult to deal with the nuclear safety issue effectively in a public debate. The "anti-nukes" can readily raise doubts about nuclear safety. It is much harder for the proponents to dispel these doubts. As Senator Pastore observed, it is easy to scare people, but very difficult to "unscare" them. Fortunately, the safety record of nuclear power is impressive. And the pros and cons are not evenly matched. The overwhelming majority of the technical community believes that the risks of a nuclear accident are much more remote than other risks which our society readily accepts.

Even though the nuclear industry has a credibility problem, that is no reason to stay out of the nuclear debate. Recognition of the credibility problem does, however, impose an essential constraint: industry representatives must be careful not to overstate their case, regardless of how tempting this may be when confronted by the overstatements of the opposition.

By actively participating in the debate, the proponents of nuclear power can avoid losing by default. However, it is unlikely that they can win on their own. Because the issue must finally be decided by the political process, political leadership is essential. The public controversy will only recede when political leaders whom the public trusts endorse the conclusion that the benefits of nuclear power clearly outweigh the risks. Whether effective political leadership will be forthcoming is uncertain. As the public controversy mounts, the number of politicians willing to speak out in favor of nuclear power, not surprisingly, has decreased. However, during a period when electrical demand is not growing, there is no clear need to bring the nuclear debate to an end. The political instinct not to take on a controversial issue prematurely is certainly understandable. We may hope that when the need to make difficult choices becomes more compelling, the necessary leadership will be provided.[8]

IMPROVING THE REGULATORY PROCESS

A reform of the nuclear regulatory process, to reduce the element of uncertainty and the time required to obtain approvals, would be highly desirable. I suspect, however, that substantial regulatory reform will not be possible until the public acceptance problem is resolved. The principal cause of delays and uncertainties is that the regulatory process is being used as a political battleground by the opponents of nuclear power. It is easy for lawyers to obstruct and delay regulatory proceedings. Changing NRC's procedures cannot eliminate that problem. The only real solution lies in resolving the underlying political controversy. Only then can the regulatory process return to its proper objective of providing a careful technical review of the safety features of proposed plants.

Even though the regulatory process will remain a political battleground for some time to come, the development of increased public confidence in the regulators is essential. Nuclear safety reviews inevitably involve complex technical issues. Because the general public cannot participate effectively in deciding the underlying issues—no matter how "open" the process becomes—trust in the good judgment and integrity of the regulators is essential. The establishment of the Nuclear Regulatory Commission as an independent agency was an important step in the right direction. NRC's efforts to make the regulatory process more open was desirable in helping to dispel an atmosphere of secrecy and suspicion.

It is important to the future of a workable regulatory program that the principle of federal preemption be reaffirmed. While this principle was clearly established by the Atomic Energy Act, in recent years it has been chipped away by an increasing number of state actions.[9] The real problem once again is one of public confidence. Only when the political leadership at the state and local level concludes that the general public is willing to trust the determinations of a federal regulatory agency will the pressure to interpose state and local restrictions decline.

---

[8] The vote on Price-Anderson extension in the fall of 1975 suggests that when an issue is recognized as important to the viability of nuclear power, substantial congressional majorities will take a pronuclear stand.

[9] For a discussion of federal preemption, see Chapter 4.

THE NUCLEAR FUEL CYCLE

During the past two decades the principal focus of the nuclear effort has been on the development of nuclear power plants. The nuclear fuel cycle was in large measure taken for granted. The front-end of the fuel cycle—uranium mining and enrichment—had been developed on a large scale, in the 1940s and 1950s, to meet the demands of the nuclear weapons program. With the decline in weapons production, there was ample capacity to serve the slowly growing needs of the power program. As for the back-end of the fuel cycle—spent fuel reprocessing, plutonium fabrication, and waste storage had all been conducted on a large scale in conjunction with government programs. The general assumption was that the private sector would proceed to build whatever fuel cycle capacity was necessary when required for the growth of nuclear power. This was a popular assumption. It fitted the conventional economic wisdom of both government and industry leaders and it did not require the appropriation of government funds.

Any serious consideration of how to proceed with the nuclear option demands a more realistic look at the nuclear fuel cycle. Question about the adequacy of enrichment capacity, the lack of reprocessing, and the uncertainty about long-term waste storage are of real concern to utilities which must look three and four decades ahead when they themselves commit to building a nuclear plant. In addition, the unresolved problems of the fuel cycle affect the public acceptance issue. The antinuclear campaign has specifically focused on plutonium use and on the lack of reprocessing and waste storage capacity. Because the problems differ, each segment of the fuel cycle must be considered separately.

Uranium Supply—The rapid rise in uranium prices following the Arab oil embargo caused widespread concern. However, the present uranium problem appears to involve primarily a temporary disarray of supply and demand conditions.[10] Demand grew very rapidly as a result of the large wave of nuclear orders in the early 1970s. Supplies did not keep pace because of the long lead time required to develop new mines, and because large foreign supplies were kept out of the market. The U.S. mining industry has substantial undeveloped reserves and rapid price increases provide a strong incentive for opening new mines and expanding exploration. The best evidence available suggests that estimated reserves are

---

[10] By suggesting that we are faced with a relatively short-term disarray of supply and demand conditions, I do not mean to imply that uranium prices are likely to return to the levels of the early 1970s. My only point is that concern about a uranium shortage is likely to prove only a short-term problem.

adequate to cover the full lifetime fuel requirements of all of the nuclear plants which are likely to be built during the next two decades.

If the nuclear option is pursued aggressively, we may well face a uranium supply problem in the longer term future. However, within the next two decades there should be ample time to deal with that problem. The development of breeder reactors would, in effect, extend our uranium supplies practically one hundred fold. Furthermore, additional uranium reserves are likely to be discovered and new mining techniques may be developed.

Uranium Enrichment—In 1976 the U.S. enrichment capacity consisted of the three government-owned plants built in the 1940s and 1950s to supply the needs of the nuclear weapons program. These plants were being upgraded and expanded and were capable of meeting the antici- pated demand for power reactor fuel until well into the 1980s. It was originally expected that new enrichment capacity would be needed as early as 1983. With the slowdown in nuclear plant construction, a new plant may not be required until the mid-1980s. However, the long lead time required to launch an enrichment project means that the debate whether the government or private industry should build the next plant cannot continue much longer.

Both the Nixon and Ford Administrations strongly encouraged private industry to build enrichment plants. This policy was reflected in the Nuclear Fuel Assistance Bill introduced by the Ford Administration in 1975. A short review of the economic and competitive dimensions of uranium enrichment raises doubts about the realism of this policy. On any normal risk/reward analysis, uranium enrichment does not look like an attractive opportunity for private participation.

First of all there is the magnitude of the investment required. To build a gaseous diffusion plant, still the only "proven" technology, re- quires an investment in the $3 to 5 billion range. While a plant using centrifuge technology would be somewhat less costly, it would probably require an investment in the $1 to 3 billion range, depending on the size of the plant.

Even more troublesome is the long negative cash flow cycle. Figuring roughly, a company considering the enrichment business must face an initial decade in which several billion dollars must be invested for plant construction before enriched uranium is produced. There will then follow a decade or two before the initial investment can be recouped. Thus, it may well be the end of the twentieth century before the potential entrant can expect to be "in the black." Thus, the ultimate profitability of his investment will depend on the price levels and permissible return on investment, which can be earned in the twenty-first century. Investing several billion dollars in the next decade, in the hope of earning large profits in the twenty-first century, does not look like a very tempting proposition. It would be prudent to assume that enrichment prices may be subject to continuing government control. What the regulators of the twenty-first century will consider a proper return on investment represents one more element of uncertainty.

Before worrying about the twenty-first century, the potential entrant into the enrichment business must recognize that the price level charged by his new plant may be compared with the prices charged by the existing government-owned plants. These were built at a time of much lower construction costs and can charge practically any price which, as a matter of political judgment, appears desirable.[11] In addition, the new plant will face competition on the world market from a number of plants controlled by foreign governments. Furthermore, from a technical standpoint, the new entrant must consider whether his plant could be made obsolete by new developments in laser separation technology or by further advances in centrifuge technology.

Finally, there is an element of unreality in looking at the enrichment business as if it were an ordinary business investment opportunity. Concerns about the proliferation problem and about safeguarding nuclear materials against diversion and sabotage add an overriding element of political risk. There is no way of calculating realistically what types of restrictions and what additional costs will be imposed by national or international policies directed at the proliferation and diversion problems.

One of the desirable elements of flexibility in our economic system is that there is no rigid demarcation of the proper areas for private initiative. However, regardless of where one chooses to draw the line, it is hard to conclude that uranium enrichment falls on the right side for private investment. It has been proposed that the principle of encouraging private initiative be maintained through governmental assumption of the risks which would otherwise discourage private participation. To me this represents a quixotic exercise to overcome reality in dogged adherence to an ideological abstraction. For the foreseeable future it would be far

---

11 The government discretion can be exercised in either direction, for example, it has been proposed that the prices charged by the three government plants should be raised substantially in order to encourage private entry into the enrichment field. This illustrates the complexity of private entry into a field in which the government plays such an all-pervasive role.

more realistic to accept the fact that the political risks, and particularly the concern about proliferation, are sufficiently great to make it desirable that the government build the next increments of enrichment capacity.

Spent Fuel Reprocessing—During the second period of nuclear enthusiasm, which accompanied the initial success of the turnkey orders in the mid-1960s, several private entries were made into the fuel reprocessing field. Three plants were actually constructed. However none is operating today. The Nuclear Fuel Services plant in upstate New York was shut down after several years operation, and rebuilding plans are under consideration. General Electric's plant in Morris, Illinois, never succeeded in becoming operational. The Barnwell, South Carolina plant built by Allied-Chemical and Gulf Oil has been largely completed but is unlikely to be put into operation in the foreseeable future.

At present, the major obstacle to progress in the reprocessing field is the absence of government decisions on several key issues. These decisions are needed to enable private industry to decide whether and on what basis to proceed. The issues involved are, of course, highly controversial and it is not surprising that they have proved hard to resolve.

The issue of plutonium recycle involves not merely the fundamental question whether plutonium should be used as a power reactor fuel, but also what restrictions should be applied to its use as a fuel, and to its processing and transportation. Both health and safety issues and safeguards considerations are involved. Until these questions have been resolved, it is impossible to make a meaningful analysis of the economics of reprocessing. One cannot accurately estimate either the costs of reprocessing, or the value of the end products. Thus, basic "go or no go" decisions cannot be made.

Equally important to the future of reprocessing are the still unresolved government decisions regarding waste storage. The form in which radioactive wastes will be stored, and whether the reprocessor or the government must bear the cost of turning the wastes into their final form for storage, can have a significant impact on the economics of reprocessing. For example, if high level wastes must be vitrified before they can be turned over to the government for storage, an additional step has to be added to the reprocessing cycle. It has been estimated that for the Barnwell plant, the cost of adding the additional step could exceed $100 million.

The volume of spent fuel which will be produced during the next decade is not of a magnitude which makes prompt action essential. The capacity of existing storage facilities can readily be increased several fold. Furthermore, additional storage basins can be constructed at a cost which, in terms of fuel cycle economics, is relatively modest. Thus, there is no need to rush to immediate decisions. It is far more important to reach sound long-term conclusions.

Whether additional reprocessing plants should be built by private industry or the government is far from clear. The estimated cost of building a reprocessing plant is in the $200 to 500 million range. Such amounts are not beyond the level required for other chemical plants built by private industry. However, it is clear that reprocessing, like enrichment, involves overriding political issues, including proliferation policy and safeguards against diversion of nuclear materials. For example, it has been proposed that reprocessing plants should be located on government-owned and controlled sites. I question whether reprocessing should be considered a private sector activity in the foreseeable future.

The experience with the private enrichment plants indicates that there are development problems in making the transition from government reprocessing technology, using low burn-up metal fuels to commercial reprocessing of high burn-up uranium oxide fuels. Therefore, it would be desirable for ERDA to take the steps necessary to get several reprocessing plants into operation. The results obtained from the operation of these plants will provide a better basis for deciding whether subsequent plants should be built by the private sector or by the government.

Radioactive Waste Disposal—Large volumes of high level waste products requiring final disposal will not build up until after spent fuel reprocessing plants are in operation. Thus, from the standpoint of physical need, there is no urgency to proceed with the construction of waste storage facilities. However, the lack of decisions in this area has become a significant issue in the nuclear debate. For that reason it would be highly desirable to take steps to remove from the nuclear debate an issue whose importance has been exaggerated far beyond its real dimensions. Extensive experience under government programs shows that the construction and operation of waste storage facilities present no insurmountable technical problems. There also is adequate geological data on potential storage sites. There have been innumerable studies and the time has come to decide on the form of storage and to choose a location.

It would seem beyond argument that long-term waste storage must be a government activity. The time periods for storage run into centuries. In addition, the health and safety concerns call for extensive physical isolation and effective protection against theft or sabotage.

DEVELOPING REALISTIC BUSINESS APPROACHES

There is an Alice-in-Wonderland quality in the oft-expressed concern of the critics of nuclear power over the huge profits of the nuclear industry. As the historical overview indicates, nuclear energy has never become a satisfactory business. An industry which has gone through such severe cycles cannot remain very healthy.

Relations between the utility industry and the reactor manufacturers have often been difficult and the present climate is disturbing. Turnkey contracting was a disaster for the manufacturers. However, the current approach, under which reactors and fuel are sold as if they were items of conventional hardware, is also unsatisfactory. New approaches must be developed which improve the opportunities for overall systems integration, and which provide a more flexible framework for dealing with the uncertainties of the licensing process, the need for technical evolution and the other unpredictable factors which can change and delay nuclear projects. To structure nuclear contracts as if the risks were the same as in contracts for turbine generators represents an exercise in make-believe which, in the long run, hurts both the utilities and the manufacturers.

Unless new business approaches are developed which face up to the real problems of the nuclear business the future of private participation is not encouraging. It is clear that the romance has gone out of the business. No company will enter the nuclear business, or remain in it, because Wall Street will give its stock a higher price-earnings multiple. Either the business will be set up on a basis which provides a favorable balance of risks and rewards or there is no sound basis for continued private participation.

FINANCING DEVELOPMENT WORK

It is also desirable to face up to the need for more development work to enable the light-water reactors to achieve higher on-line availability. Improved availability is essential to the realization of the economic benefits of nuclear power. The advantages to electrical consumers in terms of lower generating costs would be very sizable. Further development testing to prove out the performance of reactor safety systems would be desirable both on objective technical grounds and in order to overcome the public concerns created by the nuclear critics.

In both of these areas government support is needed. Considering the present economic condition of the nuclear industry, it would be unrealistic to expect industry funding of large new development programs. To withhold government support on the theory that water reactors are a "commercial product," and that government spending would represent an improper subsidy, is to put a real problem into an ideological straight-jacket. The proper areas for government funding can only be defined on pragmatic grounds. To say that government funding of breeders is appropriate because they have not been "commercialized," while support of water reactors is inappropriate, makes no practical sense. The economic rationale for the breeder—extending the available uranium supply —will become meaningful only if there is an expanding light-water reactor economy. If there is not, the need to extend existing uranium supplies will never arise.

In other major industrialized nations, government support for long time-cycle, high technology businesses is provided with the avowed objective of strengthening private industry. To explain to a Japanese or a German, not to mention a Russian, that the U.S. government is reluctant to support water reactor development because water reactors are "commercial products," produces stares of incredulity and incomprehension.

Conclusion

We are now in the third "ice age" in the brief history of the nuclear business. In such a period it is much easier to look at the failures than the successes. The technical successes have been enormous. We now have in operation nuclear power plants with a total generating capacity greater than the total U.S. electrical capacity installed in the year 1940. Nuclear power has demonstrated a significant economic advantage over coal and oil, and has maintained a superb safety record. These are real accomplishments. The U.S.-developed light-water reactor technology has become the standard of the world. Japan, Germany, France, Italy, Spain, Sweden, and Switzerland have all adopted our water reactors for their nuclear programs. The technical problems which have been solved are much greater than those remaining to be solved.

Our principal problems do not lie in the technical area but in the political area. We are caught in a period of political stalemate and national irresolution. Unless there is strong political leadership to proceed with the nuclear option, the opponents will fill the policy vacuum.

The steps to be taken to get out of the present miasma are clear. First, there must be strong affirmation of the nuclear option as a key part of U.S. energy policy. This must be accompanied by a program to deal forthrightly with the public concerns about nuclear safety. The public must be reassured that the nay-sayers are a tiny minority and that the overwhelming technical judgment supports the safety of nuclear power. Second, government action should be taken to provide additional enrichment capacity, to resolve the uncertainties besetting the back-end of the nuclear fuel cycle, and to strengthen development support for water reactors. Third, the nuclear industry must develop more realistic methods of doing business.

If these steps are taken, the utilities and the nuclear industry can proceed with the construction of a sufficient number of nuclear power plants to provide the country with a properly balanced energy supply.

The consequences of failure to act are serious. The nuclear industry has little margin left for absorbing further shocks. If the deterioration of 1974-76 continues, our industrial nuclear capability might well erode beyond recovery by 1980. The effects on the U.S. economy, in terms of economic growth and energy costs, energy independence and balance of payments, as well as environmental effects, may not become apparent until the mid-1980s. By then it will be much too late to reverse directions.

Arthur W. Murphy

4

Nuclear Power Plant Regulation

In another field a description of the regulatory setting in which activity must be carried on might be a necessary part of the background for discussion. In the case of nuclear power, government regulation is so pervasive that the regulatory process is itself an important element of the controversy. Although NRC regulations cover all operations in the fuel cycle, I have confined my attention to the regulation of nuclear power plants, surely the heart of the regulatory process to date. In doing so, I do not mean to suggest that safety in other operations is not important; but, by and large, except for problems of security if we go ahead with plutonium recycle, the operations do not seem to pose issues unique to nuclear energy. Nuclear power plants do pose such issues. I will try briefly to describe the process, to identify those aspects which are most controversial, and to touch on some recent changes and proposals for change which seem worthy of note. Finally, although it is not technically a "regulatory" measure, I will describe the much discussed and little understood Price-Anderson Act.

Allocation of Regulatory Responsibility

THE ATOMIC ENERGY COMMISSION—NUCLEAR REGULATORY COMMISSION

The Atomic Energy Act of 1954 ("the Act"), established the basic scheme of federal regulation of atomic energy which is still in effect. The beginnings of our nuclear energy program were, of course, in the "Manhattan Project," created for the development of atomic bombs in World War II. Initially, the project, and indeed the whole field of atomic energy, was a highly secret activity monopolized by the federal government. The Atomic Energy Act of 1946 transferred the control of the development of atomic energy to a civilian agency, the Atomic Energy Commission (AEC). However, as the federal government retained the ownership of all fissionable materials and related facilities, and private activity was restricted to contractual operations for the government, the monopoly persisted.

The Atomic Energy Act of 1954 ended the government monopoly over atomic energy and opened the door to participation by private industry in the development of the peaceful uses of this new technology under a comprehensive statutory program of federal licensing and regulation. Resting its regulation of the use and development of atomic energy on its constitutional power to provide for the common defense, to regulate interstate and foreign commerce, and to control United States property, Congress declared that the purpose of the act was to encourage widespread participation and maximum scientific and industrial progress in the development and utilization of atomic energy for peaceful purposes.

The act established an elaborate system of licenses and regulation to control private participation. No one could build a reactor, possess nuclear fuel, or operate a nuclear power plant without a license from the AEC. And even after the licenses were secured, operation was (and still is) subject to the control of the AEC (now the Nuclear Regulatory Commission).

Although no one has ever seriously questioned the need for extensive regulation of the nuclear power industry, the administration of the regulatory program has been the subject of almost constant controversy for the twenty-odd years of its existence. Those subject to regulation complained that the regulation was oppressive; others that it was inadequate. There has been at least some justice in both positions.

Two of the most severe criticisms of the AEC over the years were: (1) the combination in one agency of the functions of promoting the development of nuclear power, and regulating that development, and (2) the refusal of the AEC to concern itself with nonradiological impacts on the environment, chiefly "thermal pollution" by discharges of heated water into lakes and rivers. Both of these criticisms may have been muted by recent events. The enactment of the National Environmental Policy Act (NEPA) in 1969 settled the question of AEC responsibility to consider nonradiological insults to the environment. More recently, the Energy Reorganization Act of 1974 split the AEC's jurisdiction between the Energy Research and Development Agency (ERDA) and the Nuclear Regulatory Commission (NRC). The former was given the responsibility —*inter alia*—for development programs, whereas the latter was given the regulatory functions. It is much too early to tell whether the separation of functions will have the magic effect its proponents expect of it, or whether, in the long run, the divorce of the regulatory personnel from involvement in new technology will make regulation less effective—as some opponents of separation maintained. In any event, our review of the regulatory picture will take as its starting point the system as it now exists, with NRC having primary responsibility for radioactive effects but required to consider environmental effects of other kinds.

THE ROLES OF THE AGENCIES

Under the current provisions of the Atomic Energy Act, the regulation of general radiation hazards is shared by the NRC, several other federal agencies, and state governments. However, nuclear power reactors are regulated almost exclusively by the NRC, although that monopoly is being challenged on several fronts. As this was being written, there was pending in the United States Supreme Court, a case involving the allocation of authority between NRC and the United States Environmental Protection Agency (EPA) for setting standards governing discharges from nuclear power plants. Of even greater potential significance are the various measures—bills and initiatives—pending in many states which seek to stop development of nuclear power. The first of these measures (an initiative) was scheduled to be voted on in California on June 8, 1976. Accordingly, it seems useful to spend a little time discussing the present allocation of responsibility within the federal government and between the federal government and the states. I do so even though, in my view, the state moratorium measures are almost certainly unconstitutional as presently written.

OTHER FEDERAL AGENCIES

NRC has the basic responsibility for regulating radiation hazards from by-product, source, and special nuclear materials. It has established criteria governing permissible doses, levels, and concentrations of radiation and precautionary procedures. NRC has also established rules governing the licensing of nuclear materials. Other federal agencies and departments, however, do have some regulatory responsibilities over these materials. Although NRC has issued rules establishing requirements for the transportation and packaging of licensed materials, licensees must comply with the regulations of the Department of Transportation and of the Postal Service which are appropriate to the mode of transportation. EPA has been given authority to establish generally applicable environmental standards for the protection of the general environment from radioactive material and to regulate the disposal of wastes into the oceans under a permit system.

The only qualifications on the exclusive authority of NRC to regulate the construction and operation of nuclear power plants have resulted from recent environmental legislation; however, the impact has, to date at least, been minimal. Although, as noted above, EPA has been given the authority to establish generally applicable environmental standards for radiation protection, both EPA and NRC have construed this authority as limited to the establishment of ambient standards and not extending to the imposition of restrictions on discharges from individual licensed facilities.

The situation under the Federal Water Pollution Control Act Amendments of 1972 (FWPCA) is potentially very different. Pursuant to the FWPCA, EPA has regulatory authority over, among other things, discharges of certain pollutants (for example, thermal pollution) from nuclear power plants into navigable waters, and EPA and NRC have established an agreement to coordinate NRC's responsibility under NEPA with respect to the environmental impact of the same discharge of pollutants. The agreement, however, specifically excludes effluent limitations on source, by-product, and special nuclear materials as subject to regulation by the commission, and the administrator of EPA has excluded these radioactive materials from the definition of "pollutants" within that agency's jurisdiction. Notwithstanding clear legislative history supporting the EPA and NRC interpretation of the FWPCA to give the NRC exclusive responsibility for radioactive effluents from nuclear power plants, the Court of Appeals for the Tenth Circuit has recently held that the EPA must assume responsibility for regulating the discharge of all radioactive materials, including source, by-product, and special nuclear materials, into navigable waters. The consequences of this decision, if it is not reversed (it is now before the Supreme Court), could be far-reaching. At the least it would seem to impose a dual (EPA and NRC) federal responsibility for the radioactive liquid effluents of nuclear power plants. Moreover, given the emphasis of the FWPCA upon the "primary responsibilities and rights of the states to prevent, reduce, and eliminate pollution," an affirmance by the Supreme Court could arguably open the door to some state regulation.

THE ROLE OF THE STATES

Pursuant to section 274(b) of the act, the commission is authorized to enter into an agreement with the governor of any state and to discontinue its regulatory authority over by-product, source, and special nuclear materials in quantities not sufficient to form a critical mass. For the duration of such an agreement the state has full authority to regulate, within the state, the materials covered by the agreement for the protection of public health and safety from radiation hazards.

Section 274(c) provides that the commission shall retain authority and responsibility with respect to the regulation of (1) the construction and operation of any production or utilization facility; (2) the export or import of by-product, source, or special nuclear materials and of any production or utilization facility; (3) the ocean disposal of by-product, source, or special nuclear waste materials; and (4) the disposal of other by-product, source or special nuclear materials determined by the commission to represent a special hazard. However, with respect to any application for a commission license authorizing such an activity, the commission is required to give prompt notice to the state in which the activity will be conducted and to permit the state to offer evidence, interrogate witnesses, and offer advice on the application.

Until very recently, most states accepted the exclusivity of federal control over the safety of nuclear power plants. Some states attempted to impose their own regulation—a notable example being the effort of the Minnesota Pollution Control Board to impose more stringent requirements on discharges from nuclear power plants than were imposed by the AEC, which led to the decision in *Northern States Power Co.* v. *AEC* [1] that states be preempted from the field. Lately, however, opponents of nuclear power have made a major effort to enact legislation in the states. Although there are variations in the details of these proposals, their common theme is a provision for a "moratorium" or rollback of construction or operation of existing plants unless one or more of the following conditions is satisfied: nuclear power plants are demonstrated to be safe; nuclear wastes can be stored safely forever; adequate safeguards against diversion of nuclear materials (for example, by terrorist groups) are devised; the utility operating the plant waives the limit of liability imposed by the Price-Anderson Act. (The provisions of the Price-Anderson Act are discussed later.) In my view, if such proposals are enacted they will almost certainly be declared invalid because the area of regulation is preempted by Congress. However, the states do retain authority to regulate nonradiation aspects of power production, for example, rates, and their influence will undoubtedly be felt. Moreover, there is some

[1] 447 F.2d 1143 (8th Cir. 1971) *aff'd per curiam* 405 U.S. 1035 (1972).

sentiment for a reallocation of some responsibilities, and the picture may change in the relatively near future.

The Licensing of a Reactor

THE AGENCY REVIEW PROCESS

The licensing of a nuclear reactor is accomplished in two distinct phases: before the applicant can begin construction of the reactor he must receive a "construction permit;" after construction is completed he must apply for and receive an operating license. The application for a construction permit is evaluated in great depth by the staff of the NRC. The staff analysis covers all aspects of the proposed reactor, including the site of the proposed reactor from the point of view of location, meteorology, geology, hydrology, seismicity, the design of the reactor, the adequacy of auxiliary plant systems, the quality control system, the technical qualifications of the applicant, and the environmental impact. In the course of staff review other agencies of the federal government are usually invited to express their opinions on the project within their respective areas of expertise. The Geological Survey of the Department of the Interior may be called upon to review geology and hydrology; the Coast and Geodetic Survey of the Department of Commerce may furnish a study of seismicity of the area; and the Weather Bureau in the Department of Commerce (now incorporated with the Coast and Geodetic Survey and other units into the Environmental Sciences Administration) may be requested to make a report on meteorology. The staff may also call upon consultants, inside or outside the government, to evaluate specific aspects of reactor safety.

In addition to review by the staff, all applications for construction permits are required, under existing law, to be submitted to the Advisory Committee on Reactor Safeguards (ACRS). The ACRS, originally established by the AEC *sua sponte,* became a statutory body in 1957. It consists of fifteen members (serving on a part-time basis) representing the various disciplines involved in an evaluation of reactor safety. The ACRS reviews the application, consults with the applicant and the staff, and reports to the commission its conclusion as to whether the proposed reactor "may be operated without undue risk to the public." The ACRS report is conclusive in form although it may often specify areas of study which it believes must be looked into before starting operations.

The act requires that a public hearing be held before the grant of a construction permit. This hearing—which is mandated whether or not requested by an interested person—was originally held before a hearing examiner.[2] In 1962, as part of a general reshaping of the regulatory process, a new entity was authorized: *ad hoc* safety and licensing boards consisting of two "technically qualified" members and one member "qualified in the conduct of administrative proceedings." The board's mandate is to find out whether the facility may be constructed without undue risk to the health or safety of the public. The decision by the board is subject to review by the commission. In the last few years, most of the appellate function of the AEC was exercised by an Atomic Safety and Licensing Appeal Board (ASLAB). The ASLAB was created with a view to further insulating the commission's promotional function from its licensing function—an objective no longer meaningful in the case of NRC. It is not clear how the ASLAB and the commission will allocate their functions. In any event the agency's decision is also subject to court review.

Between the start of construction and the beginning of operation, the staff evaluation is more or less continuous. Before an operating license is issued the ACRS must again pass a ruling on the application. Since 1962 there has been no requirement for a hearing at the operating license stage except on the request of a party. NRC regulation continues throughout the life of the reactor.

THE SAFETY ASSESSMENT

Beyond question, the aspect of nuclear power plant licensing which gives it its unique character is the possibility, remote though it is generally conceded to be, of an accident causing substantial damage to persons and property off-site. Whether one agrees or disagrees with the view that radiological discharges in the course of normal operation are not a serious problem; and whatever one's view of the process of considering nonradiological effects on the environment,[3] it seems clear that if normal discharges and such environmental effects were all we worried about, the regulatory process would be vastly different.

[2] Until about 1968 most hearings were "uncontested." Since then the norm has been a contested hearing with the opposition coming from an intervenor or group of intervenors.

[3] This description focuses on the radiological questions. As noted earlier NRC must also consider nonradiological environmental impact.

THE BASIC APPROACH

To some extent the problem of safety at a nuclear power plant is no different from that at any other plant. In many respects the basic plumbing is the same—standards for pressure vessels, pipes, etc., are already etsablished; and experience with valves, engines, etc. can be relied on. In other respects, however, the technology is new, and experience in existing reactors, although comforting, has not been long enough to test many systems in actual operation. In another field, the absence of operational verification would probably not cause much concern. The traditional approach has been to learn by experience. In the nuclear field, however, the possible consequences of a serious accident are such that the traditional approach is not adequate. As a result, probably to a unique extent, the field of reactor technology has been marked by attention to what might happen. In a very real sense, the accident that people worry about most is the one no one has thought of. The typical pattern is for the ACRS to suggest a possible problem, for the staff to require a solution and the applicant to propose one. As to those which have been anticipated, the basic approach has been to provide backup systems to deal with the event should it occur, and to provide "redundant" systems where thought necessary. For example, since many systems are dependent on power, what will happen if the local power supply fails? The response is to provide a backup diesel engine. Suppose the diesel fails? The response is a second diesel. The result is that, as is indicated in Chapter 1, for a serious accident to occur, a great many things have to go wrong at the same time. But the question remains: what is an acceptable risk?

THE ANALYSIS OF RISK

Radioactivity can be dangerous, and the only way in which one can be absolutely sure that no damage is done is not to build reactors—a course of action which so far has been rejected. Granted then that some risk will be taken, how much will be tolerated? Logically this "policy" decision should precede the work of the regulatory staff, the ACRS, and the boards. The act itself contains no guidance except the injunction in Sections 103 and 104 that no license is to be issued to a person if in the opinion of the commission the issuance "would be inimical to . . . the health and safety of the public." The NRC regulations bearing on the problem are contained in 10 CFR Part 100, "Reactor Site Criteria." Part 100 contains the general admonition that:

> It is expected that reactors will reflect through their design, construction and operation an extremely low probability for accidents that could result in release of significant quantities of radioactive fission products. In addition, the site location and the engineered features included as safeguards against hazardous consequences of an accident, should one occur, should insure a low risk of public exposure.

After this admonition, Part 100 goes on to list a number of specific factors to be taken into account, including population density and such physical characteristics of the site as seismology, meteorology, and so forth. Population density is a critical element. For example, the regulations contemplate an "exclusion area" around the reactor, and a zone around the exclusion area in which the population density is sufficiently low that the residents could be evacuated or otherwise protected in case of a serious accident. It is expressly recognized that all of these criteria (except possibly that requiring the site to be no closer than a quarter mile from an active earthquake fault) are guidelines rather than prerequisites to licensing, and, indeed, the regulations state specifically that, notwithstanding unfavorable physical characteristics, a site may be acceptable "if the design of the facility includes appropriate and adequately compensating engineering safeguards." It is this authorization to rely on engineering safeguards that opens the way to urban siting even within the present criteria.

Obviously these regulations do not answer the question as to what risk will be tolerated—except to indicate that the tolerance should be low. Given this general "standard," the regulation might have proceeded to quantify the risk in terms of the magnitude of exposure and the likelihood of releases of radioactive materials. So far this has not been done. Instead, the practice has grown up of measuring the safety of design by its ability to withstand an accident of postulated intensity. The genesis of this practice is, again, in Part 100. For purposes of calculating the necessary "exclusion area," "low population zone," and "population center distance," the applicant is required to calculate the exposure from an accident involving a meltdown of the core and substantial fission product release. The accident postulated is directed to be one resulting in potential hazards "not exceeded by those from any accident considered credible." The regulation does not elaborate any criteria for the finding of credibility, and it is obviously an imprecise standard.

Proposals for Change

This elaborate process pleases no one. Industry has a number of complaints about regulators. In general, the theory of "defense in depth" is accepted, but some think that "redundancy" can be carried too far, especially where it requires expensive "back fitting" of previously licensed reactors. But the chief objection is to the inordinate length of time which the process takes; although the NRC review is not the only administrative hurdle which a power plant must surmount (as many as thirty-five separate permits may be required from various local, state and federal agencies) it is certainly the major one. It now takes some nine to ten years for a nuclear power plant to come on line after plans are drawn. Most observers believe that this is about twice what it should take and that it should be possible to bring a plant on line in about five years. As is discussed in chapter 2, the cost of the stretch-out is enormous. All of the stretch-out cannot be attributed to administrative delays; construction delays are a major factor. But concern with the length of the process is certainly justified. Indeed we may have reached the point where the length of the process compared to fossil fuel plants may dictate the choice of plant. And in any event the cost of borrowing is such that delay may mean that no plant can be built.

On the other hand, critics of the process charge that however long it may take, the process does not produce safe reactors. Their criticism takes many forms: the failure of AEC-NRC to provide a permanent solution to the waste storage problem; the failure of the regulatory process to address itself to dangers of sabotage or nuclear diversion; the failure to perform basic research before licensing reactors; the inadequacy of quality assurance programs (compared, for example to the naval reactor program). Others could be cited, but essentially they are aspects of the pervasive criticism that safety is sacrificed to economics. This criticism, although somewhat muted in the early days of the NRC, has once again become a focal point of the opposition to nuclear power. (It will be considered at some length after a discussion of some proposed changes in the process.)

Spurred by a presidential directive, among other stimuli, that the licensing process be substantially shortened (to between five and six years) a number of suggestions for change have been put forward in recent years. They include: elimination of the hearing entirely ("at best a charade, at worst a sham"); [4] elimination of the mandatory hearing for a construction permit; elimination of the hearing at the operating license stage; advance approval of site, that is, before plans for a particular plant are formulated; abolition of the mandatory ACRS review except for new types of reactors; enlargement of the type of prepermit site work allowed; use of "generic" hearings—that is, a hearing devoted to an issue (or issues) common to a number of individual proceedings, for example, the emergency core cooling systems of water reactors—and the concomitant elimination of those issues from individual licensing hearings; increased standardization of plant types, licensing of standard types, and restriction of the hearing to consideration of site-related issues; provision for interim licenses; elimination, or substantial alteration of the Department of Justice "antitrust review;" and joint proceedings for federal and state licensing bodies.

Some important changes have already been effected. NRC has recently completed an exhaustive self-analysis and promulgated detailed instructions for standardizing review procedures. Effectuation of other proposed changes could have a substantial effect on the length of the licensing process; whether they are politically feasible may be another question. However desirable it might be in theory (I would not favor it) the time does not appear right for the elimination of the hearing. Elimination of the requirement of a hearing at the construction permit stage, even if no party requests it, would probably be acceptable, but at this time, when intervention is the norm, it might have limited utility. On the other hand, elimination of the hearing at the operating license stage would certainly save time and might be acceptable now that technology is more stable.

Proposals for "generic" hearings and standardization have much to recommend them. There is no justification for relitigating the same issues over and over again. However, care must be taken that they do not destroy the opportunity for public participation in the hearing process. (See *The Issue of Public Participation in the Hearing Process,* below.)

Finally, the development of joint proceedings for federal and state licensing bodies should be pursued—if only to guard against further prolongation of the hearing process. Duplication between NRC and state agencies on environmental matters is already a serious problem, and it seems likely to grow more serious with increased state involvement in nuclear power plant licensing.

---

[4] This refers to the most dramatic of the proposals by Professor Harold P. Green. His proposal and many of the others are discussed in a symposium, "The Nuclear Power Plant Licensing Process," 15 W. & M.L. Rev. 487 (1974).

Acceptability of the Safety Decision

None of the proposed changes in licensing procedures promises to solve the problem of acceptability of the NRC conclusion about safety and, indeed, it is hard to be anything but pessimistic about the chances of finding a solution to that problem. The fundamental criticism of NRC is that it allows consideration of economics to override considerations of safety. But unless one means to insist on absolute safety—which means no reactors—there is always a choice to be made. If the discovery of *any* safety problem is to necessitate shutdown, nuclear power cannot be economically viable. There must be a standard of reasonableness attached to regulation.

Take the controversy over Indian Point II, for example. A major criticism of NRC by its resigned project supervisor was that it did not shut down Indian Point II when "it discovered" that some valves would be submerged and, therefore, inoperable in the event of a loss-of-coolant accident (LOCA). All agreed that the valves should be relocated; that already had been done in the case of Indian Point III and was scheduled for Indian Point II at the time of its scheduled refueling outage in spring, 1976. In view of the fact that the condition was known that there are several alternative ways of accomplishing the objective of the valve relocation on the existing reactor, and that the location of the valves only become important in the highly unlikely event of a LOCA, it seems a bit harsh to accuse NRC of "total irresponsibility" unless one is of the view that it must never make *any* trade off between safety and economics.

But how is one to judge the acceptability of a trade off? For that, it would seem that there is no alternative to frank discussion in terms of the probability of an accident, its size, and the cost of the added "ounce of prevention." Until the Rasmussen Report, the practice has been to shun talking (perhaps even to avoid thinking) in terms of acceptable risk. True, no one would say that the risk was zero, but there was always the implication that while as a man of science one could not rule out the possibility of a catastrophic accident, there *really* was no chance that one could occur. This implication, if not fostered, was certainly not negated by the use of such terms as "maximum credible accident." The result is that when tests (primitive though they were) suggest that emergency cooling water may not reach the core, or the Browns Ferry accident demonstrates that things can indeed go wrong, the NRC (or AEC before it) is put in the embarrassing position of defending nuclear safety in terms of low probability of an accident where it previously *suggested* impossibility.

It may be true—it certainly seems to be—that the chances that ECCS will
ever be needed are very remote; it also seems true that at the worst point
in the Browns Ferry accident, the reactor was still a long way from a core
melt—let alone a core melt of catastrophic consequences (see Chapter 1).
But, inevitably the agency—and the proponents of nuclear power—are
on the defensive. "If those unexpected accidents can occur, why cannot
other accidents hitherto deemed incredible also occur?" The only honest
answer is that they can, but that everything we know indicates that the
risk is so low as to be worth taking.

It may be that psychologically, the society will be unable to live with
an articulated risk of a serious accident, however remote the possibility
may be. There is at least some reason to suspect that the typical approach
to risk is to deny that it exists. But in the last analysis, the judgment
about nuclear power must be made by society, and the criteria for that
judgment must be made available.

THE ISSUE OF PUBLIC PARTICIPATION IN THE HEARING PROCESS

As noted above, until about 1968 the general public rarely became
involved in the licensing of nuclear power plants. The parties to the
process were the regulatory staff of the AEC and the applicant for a
license. Even though a public hearing was mandated at the construction
permit stage—whether or not requested by anyone—the public rarely did
more than make a "limited appearance." Frequently no one unaffiliated
with the staff or applicant appeared at the hearing except local officials
in support of the facility. In part this lack of participation was the result
of the AEC position that it did not have jurisdiction to hear nonradiological issues, since most of the time the only issues on which the public
wanted to be heard were nonradiological—chiefly "thermal pollution."

Beginning in about 1968, and especially after the enactment of NEPA
made it clear that the AEC did have nonradiological responsibilities, the
public has been very much interested in the process and insistent on an
adequate voice in agency determinations. For some, the famine became
a feast; a number of hearings became field days for intervenors and their
lawyers who sought to "try" every facet of the safety and environmental
impact of reactors.[5] Since, by and large, the intervenors did not have
the technical support or the finances to make their case affirmatively,
they tended to rely heavily on cross-examination. Whatever its merits

[5] For a description of the variety of issues raised, see Murphy, "The National Environmental Policy Act and the Licensing Process: Environmentalist Magna Carta or Agency
Coup de Grace." 72 Colum. L. Rev. 963 (1972).

may be in other fields (they are debated), cross-examination of technical witnesses about lengthy documents they have prepared can be a tedious and unproductive exercise. Moreover, the AEC staff tended to resist strongly attempts to discover background documents, staff memoranda and the like, as well as privileged material submitted by others. This latter tendency served to confirm the feeling of many that the agency had something to hide, and in any event contributed to the frustration that most intervenors felt about the process. (It also added to the length of the hearing.) Most observers and participants, even those sympathetic with the idea of public participation, agreed that, tested by the standard of contribution to the safety decision, the contribution of intervenors was minimal. A number of steps have been proposed to make such participation more useful. One, a substantial relaxation of agency resistance to requests for information has been effected. Others,[6] including provision of expert witnesses and payment of attorneys' fees remain under consideration, although the agency may need statutory authorization to pay fees even if it decides to do so.

But even the most enthusiastic supporters of public participation would probably concede that safety issues common to many or all reactors should not be litigated in every case. And in recent years, AEC and NRC have experimented with generic hearings of issues common to reactors generally. Since these proceedings are, technically, "rule-making," the hearing required is "legislative" in nature. Ordinarily in such a hearing participants are entitled only to notice and an opportunity to comment upon the proposed rules. By contrast, a licensing proceeding is regarded as "adjudication" and ordinarily participants, including intervenors, have rights typically afforded in a judicial proceeding (for example, offering testimony, cross-examination, etc.). Nevertheless, it was felt that the generic proceedings should afford procedural rights similar to those in licensing and the AEC carried on two such proceedings with varying results. The first, the "as low as practicable" hearings on standards for releases during normal operation was uneventful; the second, on ECCS, took some two years to complete, was marked by legal pyrotechnics, and was widely regarded as, at best, a mitigated disaster.[7] Recently, NRC has announced generic hearings on various aspects of the fuel cycle. Although the proceedings are described as "legislative" in

---

[6] An NRC commissioned study of questions raised by requests for financial assistance in NRC proceedings was complete in July, 1975 (NUREG 75/071).

[7] The ECCS hearing and several licensing hearings are extensively discussed in Ebbin & Kasper, *Citizen Groups and the Nuclear Power Controversy: Uses of Scientific and Technological Information* (1974).

nature, participants will be afforded substantially more procedural rights than the traditional notice and opportunity to make written comments afforded at a rulemaking hearing, but less than those in the ECCS proceeding. Even these procedures are being challenged as insufficient, however. Some compromise between full adjudicative rights and minimal rulemaking procedures seems appropriate and necessary. Whether the proposed compromise will work—or even be held legally permissible— will not be known for some time.

THE PRICE-ANDERSON ACT

The Price-Anderson Act does not bear directly on safety regulation but it is administered by the NRC. As a major talking point of the opponents of nuclear power, it certainly bears discussion.

The Price-Anderson Act was enacted in 1957, as an amendment to the still new Atomic Energy Act of 1954, to achieve two objectives. One was to protect the public against the risk of uncompensated loss resulting from the peaceful uses of atomic energy. The other was to protect industry against the risk of unlimited liability for catastrophic accidents, and thus remove a roadblock to the participation of private industry in the development of peaceful applications of nuclear energy, chiefly nuclear power, which had been a major objective of the 1954 Act.

Protection of the public was afforded by two devices. The first was a requirement that licensees maintain "financial protection" in an amount set by the AEC. For electric power reactors "having a rated capacity of 100,000 electrical kilowatts or more" the amount of financial protection is fixed under the statute as the amount of private liability insurance available.

The second device utilized was the provision of a government indemnity in the amount of $500 million to take effect above the level of financial protection required. This indemnity, like the primary insurance or other protection, covers the liability of the licensee and any other person who may be liable. The act did not create any independent right of compensation for the public. The availability to the public of the proceeds of the insurance and indemnity is conditioned upon the establishment of legal liability (of some person) under existing tort law. However, it was widely assumed that in the event of a serious nuclear accident, someone (most likely the operator of the facility) would be held liable.

For industry, protection was afforded by three devices. The first was the requirement that the financial protection furnished cover the liability of any person who might be liable—thus making unnecessary the separate provision for product liability by reactor suppliers. The second was the indemnity which gave protection of a maximum of $500 million over what was available from private insurers. The third was a limitation on liability for a single "nuclear incident" to the amount of financial protection required plus the amount of the indemnity—for most power reactors a total of $560 million.

The limitation on liability—without provision of any supplemental mode of compensation—is, of course, inconsistent with the notion of complete protection of the public. Why Congress chose to limit liability in preference to an unlimited indemnity, and how it fixed on the figure of $500 million as the amount of the indemnity, are not entirely clear. For whatever reason, it chose not to make an unlimited commitment in advance of a nuclear accident, although it made a clear promise to re-examine the amount of the indemnity if it should prove insufficient.

The Price-Anderson Act did not directly affect the law governing liability. It still does not. However, as a practical matter, the act as subsequently amended by the waiver of defense provision, achieves a system of strict liability of the nuclear industry for any covered damages up to the limit on liability. The coverage of the indemnity is quite broad. In general it extends to any damage caused as a result of a "nuclear incident" which is defined as "any occurrence . . . causing . . . damage . . . arising out of . . . the radioactive, toxic, explosive or other hazardous properties . . ." of nuclear material. The only exceptions are damages due to an act of war and damage to the reactor itself, associated property located at the site and workmen's compensation claims of employees at the site.

The effect of this combination of provisions is that with certain exceptions, the licensee of a reactor is subjected to liability up to the amount of $560 million and the public is assured of a financially responsible defendant up to that amount, but the limit of liability cuts off any further legal rights which the public might otherwise have.

Early in 1976 the act was amended in several respects. The major change is the adoption of a system of retrospective assessments payable by the licensees of electric power reactors or fuel reprocessing plants in the event of a nuclear incident causing damage above the primary level of financial protection. The amount of this so-called "deferred premium" is to be established by NRC rule within a range (prescribed in the statute) of $2 million to $5 million per operating reactor. The objective of this provision is the gradual substitution of private coverage for government indemnity and, in the case of electric power reactors, an eventual raising of the limit on liability substantially above $560 million. It is estimated that, if the predicted growth rate in the number of power reactors is realized, the government indemnity will be phased out in the early 1980s. Thereafter, as the number of reactors increases, the limit on liability will automatically rise—on the assumed growth rate it is estimated that the limit will reach one billion dollars by 1990. There is no ceiling on the level to which the limit on liability may rise.

Price-Anderson, from the beginning, has been a much misunderstood act. In the early days—and indeed, sometimes today—it was criticized as a subsidy. Clearly, at least in the traditional sense, it is not a subsidy. Nuclear industry has paid millions of dollars for insurance and indemnity protection over the last eighteen years, with negligible loss experience. Since industry is required to purchase all private insurance available, the effect is not—as is sometimes charged—to lower insurance premiums otherwise payable by industry. For industry the alternative to Price-Anderson is not more insurance but less protection.

A second argument is that the combination of insurance, indemnity, and limitation of liability removes any inducement to safety on industry's part. Considering that the plant itself (which is not covered by Price-Anderson) costs about one billion dollars only a fraction of which is covered by insurance, this argument is not very persuasive. Even if we assume (although there is no reliable empirical evidence) that the threat of liability promotes safety, there is plenty of inducement for nuclear industry to maximize safety even with limited liability.

A third argument has been that the public would be better off without Price-Anderson since persons injured by a nuclear accident could sue those responsible in accordance with the usual tort rules. It is hard to prove or disprove this proposition in the abstract, but given the vagaries of the law of liability, and the need to find a financially responsible defendant, etc., it seems highly likely that the public is better off financially with Price-Anderson than without it. However, this last argument has given way in recent years to a proposal that the devices for protecting the public under Price-Anderson be kept, but the limitation of liability be removed. This proposal has been refined in recent years by adding a requirement that the potential liability be funded, and in some of the states, that full compensation be assured.

There are several aspects to this proposal, the first of which is the protection of the public. While it seems likely that removing the limit of liability would provide a larger fund available to the public in the event of an accident, it is not certain to do so. The assessment of liability is by its nature retroactive. Even if one assumes that a financially responsible defendant can be found, there are many devices available to the courts which will enable them, if they choose, to limit the liability. There has never been a tort judgment approaching $500 million, and one may well wonder what the reaction of a court would be in the event of an accident if it were given the choice between bankrupting a major company and letting the injured members of the public absorb the cost. One cannot predict exactly how the courts will behave, but there is certainly reason to suppose that courts may shrink from imposing such liability.

Secondly, adoption of such a proposal would be a marked departure from our usual practice. Price-Anderson requires that a licensee waive substantially all defenses. Thus a California utility might be liable to the full extent of its assets for an accident in Massachusetts over which it had no control. We would, in effect, be penalizing a company for entering an industry, and while arguments for such a penalty can be made, to do so would not be consistent with our present ideas about liability.

The suggestions that the contingent liability be funded—or that full compensation be assured—betray an almost total misunderstanding of the problem. Since one cannot know the size of an accident in advance, there is no way to assure full compensation except by unlimited government indemnity. Even that—as noted below—may not work, and it has in any event been rejected. One could, at least in theory, fund some of the potential liability, but how much? The reason this problem is not amenable to solution through ordinary channels is that it involves a very small probability of a very bad accident. It does not make sense to fund against the possibility of an accident which may never happen. Either the amount will turn out to be much too much or much too little. This does not mean, however, that one should not attempt to internalize costs of accidents from nuclear power. In the long run failure to internalize may result in what turns out to be an uneconomic choice of fuels. In the short run it may result in one group of users of the same fuel profiting at the expense of others. Ideally, each activity should bear its share of the cost but no more. But the tort system will not produce ideal results, nor, in truth, will any other system of loss adjustment. There is too much slippage in the process of administration and litigation to hope for better than a gross correlation of injuries, compensation and allocation of costs. Our normal system of insurance and tort law arguably provides a gross correlation for accidents of predictably regular occurrence and a size relatively small compared to the enterprise as a whole. But the situation faced here is precisely the opposite—accidents of a probability so small that we will never build up any meaningful actuarial data but of a size large enough to distort the whole enterprise. It is extremely hard, maybe impossible, to internalize the cost of such accidents. Still, the question persists; why is Price-Anderson needed? If nuclear power is as safe as is claimed, why does industry need special protection against liability? In a sense the situation of nuclear reactors is no different than for other areas of activity. The essence of the liability problem is that we do not know, we cannot know, the maximum amount of liability which may result from a particular course of conduct. Every day, each of us as an individual does things that expose him to ruinous liability. The extent of the exposure is incalculable since we cannot know until after the last case is litigated, what the legal consequences of our actions are. The imposition of liability is necessarily "retroactive;" while we can make educated guesses at the probable consequences, we cannot know. We live with this potentially ruinous liability by taking out insurance in amounts which we hope are adequate and we trust to luck that the accident will not happen; that if it does happen, the damage will be small; or even if the damage is large, that the courts will not impose liability to the full extent of the consequences. The same is of course true of business activities. No activity is covered in the sense that the actor will surely be able to pay off all claims which may arise out of it.

The case of nuclear power plants is different only because the numbers are larger. At least most people think they are. I have never seen the scenario for the damages which might be caused if a fully loaded 500,000 ton supertanker broke up on one of our coasts—a type of accident which seems a great deal more probable than the postulated maximum accident at a reactor. But conceding the differences in degree, the essential nature of the problem is the same. Unless we fix in advance the maximum liability which can be imposed because of an activity, those who engage in that activity cannot be sure that they will not be ruined.

If the nature of the problem is essentially the same, it is fair to ask why the solution should not be the same. Why should not the nuclear industry take out insurance to some "reasonable" amount and trust to luck? They might have done so; had they done so they would, as of this moment, have saved a lot of money. One reason for not advocating such a course is that such a solution considerably increases the risk to the public. A second is that it does not make sense in any situation, nuclear or nonnuclear, for a mature company to risk its existence on the possibility of a foreseeable, however remote, catastrophic accident. Blissful ignorance may be possible in other lines of endeavor—though there is evidence that other industries are waking up to the extent of their risk. But it is a luxury which has not been permitted to the nuclear industry.

It should be noted in this connection that limits of liability are recognized in other areas of private activity, for example, oil spills, government contracts, etc., and of course the traditional limit of liability afforded by separate incorporation. Separate incorporation was rejected as a solution when Price-Anderson was originally enacted because it would detract from protection of the public.

One way of accomplishing the objective of protecting the public would be to provide an unlimited government indemnity. However, even if we substituted the public treasury for private industry we could not be entirely sure of the result. Catastrophies of the hypothesized magnitude push theories of tort law and insurance and indemnification beyond the point where they are reliable. Even in the case of the government, the courts may—as they did in the Texas City disaster—refuse to assess liability. And, in the last analysis, a future Congress need not honor the obligation which the present Congress undertakes.

Finally it should be noted that most of the arguments against Price-Anderson are really arguments against development of nuclear power by private companies. The question of whether one should use private or government means of development is discussed elsewhere. For now all that should be noted is that most of the same problems would exist under a public system as exist under the private system, except that of allocating the risk between the government and industry. If we focus on the question of *whether* we will have a national program for atomic energy rather than *how*, the debate over Price-Anderson becomes largely irrelevant.

Conclusion

Although other federal agencies are to some extent involved, and despite clamor in some states for more state involvement, the job of safety regulation of nuclear power production and related activities is the almost exclusive province of the NRC. The job is not easy or popular. On the one hand, the regulated industry objects that the licensing process is too long—so long as to threaten the economic viability of its investment in nuclear power. Most observers agree that the process takes too long and that substantial time can be saved without compromising safety, and several recent steps by NRC give promise that the time spent in *staff review* will be considerably shortened. While it would seem possible to achieve comparable savings in hearing time, a shortening of the time spent in hearings may not be feasible in the present political climate.

The NRC is also subject to substantial criticism by opponents of nuclear power—their major objection being that the agency subordinates safety to economic considerations. This criticism, which has survived the divorce of regulatory and promotional responsibilities combined in the AEC, seems based in considerable part on an insistence that the agency ignore economic considerations in assessing risk, however remote the risks may be.

In the last analysis, the key question of regulation would seem to be that of public acceptability of the safety decision. If the public is unwilling to accept the very small risks of serious accidents posed by nuclear power plants, no system of regulation will suffice.

John Gorham Palfrey

5

Nuclear Exports
and Nonproliferation Strategy

Introduction

It would be foolhardy in one chapter, devoted to the international component of United States nuclear energy policies for the years ahead, to undertake to do more than concentrate on those issues which are the subject of current intensive scrutiny within and without the government. In so doing, the temptation will be resisted to recount and appraise the lively debate over what was right or wrong about U.S. international nuclear energy policies over the past twenty years.[1] Instead, the focus will be on the situation as it is now and is likely to be in the next decade, and what the U.S. has done, and might do about it.

In the international field, the central focus of concern is how the U.S. and the other advanced countries can help meet the world energy crisis by providing nuclear power (and greatly profit from it) without overwhelming the structure of international controls against diversion

[1] See Palfrey, *U.S. Nonproliferation Strategy and the Transfer of Nuclear Technology*, California Seminar on "Arms Control and Foreign Policy"—discussion paper, no. 69 (1975).

Professor of Law at Columbia University and former Dean of Columbia College, JOHN GORHAM PALFREY *has written numerous articles on atomic energy, arms control, and law and science. In 1962–66 he was a member of the Atomic Energy Commission and subsequently was a consultant to the AEC, the Joint Committee on Atomic Energy, and currently to the Arms Control and Disarmament Agency. The views expressed in this chapter are solely those of the author.*

of nuclear materials to build nuclear weapons, that has been established over the course of the last twenty years.

The issue of the spread of nuclear weapons came to a head because of three developments. The surging price of oil following the Middle East embargo in 1973 and OPEC's control of the market convinced many advanced and developing countries that, unlike the United States which possesses substantial fossil fuel reserves, they had essentially *no* option other than to concentrate on nuclear power. As a result, it is virtually certain that by the mid-1980s there will be an accumulation of large amounts of spent fuel produced by nuclear power reactors. From this "spent fuel" it is possible to recover plutonium, which then can be used either for nuclear power, or to produce nuclear explosive devices.

The reality of this danger was highlighted by the second development —the Indian nuclear explosion of May, 1974, which provided a dramatic demonstration that the development of nuclear power with only partially safeguarded facilities gives any country which possesses a reprocessing plant a derivative nuclear weapons option.

The third development of the 1970s was the maturing of nuclear power in the advanced countries, particularly in Western Europe, and the development of a lively competition in power reactors, reactor components, and other nuclear facilities, and in all probability, by the early 1980s in enriched uranium.[2]

No one can doubt that the commercial stakes are great. Through 1974, U.S. revenues from the nuclear market had amounted to $3.2 billion in reactor facilities and $700 million for "separative work" in providing low-enriched uranium fuel. Estimated revenues from facilities are projected to rise to over $1 billion annually by 1980, with another $200 million to $400 million annually for separative work. During the next 25 years estimates of nuclear power investment, world-wide, come to as high as $250 billion for facilities and $45 billion for enriched uranium.

In terms of the overseas market, the U.S. has exported sixteen nuclear power reactors (not counting large numbers of small research reactors), with twenty-eight on order. Foreign countries, in addition to domestic construction, have exported six nuclear power reactors, with twenty on order. By the end of 1975 there were 168 nuclear power reactors in operation in 19 countries, producing a total of 73,000 megawatts (electric), and by 1980, 29 countries are expected to have installed capacity of

---

[2] Some competition in the enriched supply of enriched uranium has already developed with the modest entry of the Soviet Union into the market. See Smith, "What Price Commercial Enrichment?" *Nuclear Engineering International* (July 1974), pp. 572-84.

219,000 MWe, which is about eleven times greater than that installed in 1970.[3]

There have been disturbing indications that the exploitation of the competitive market by the principal suppliers had begun to include competition over the extent of safeguards to be imposed on the purchasers.

A second consequence of the growing competitive market has been a tendency on the part of some suppliers to tie the sale of a nuclear power plant—the most profitable component—to the sale of other equipment, particularly reprocessing plants, which make it possible to acquire plutonium in a form usable in the manufacture of nuclear explosives.

With debating the merits, a body of opinion has developed that the U.S. is now in the process of losing the best of both worlds. Having gone out of its way to transfer its nuclear power technology all over the world in the interests of developing nuclear power, while also safeguarding its use, the U.S. has seemingly been losing control of the nuclear power market it had largely created, and it has also been losing its leverage over safeguards controls.[4]

The Nonproliferation Treaty

In recent years, a major focus of U.S. efforts in nuclear arms control has been the Nonproliferation Treaty (NPT), which entered into force in 1970 and was the subject of a review conference of the members in April 1975.

Perhaps the most remarkable aspect of the NPT is that it was signed in the first place, and that it has been ratified by nearly 100 countries and signed by a dozen others. Under the Treaty, just three of the five nuclear weapons states (NWS) called upon all the nonweapons states to agree not to develop nuclear weapons or explosives (Article II) which all five NWS were continuing to develop, and in addition required them to accept mandatory international safeguards by the International Atomic Energy Agency (IAEA) on their lawful peaceful activities, whether constructed indigenously or with outside assistance (Article III).

[3] These projections are based on "The Report of the Atlantic Council," *U.S. Nuclear Fields Policy* (1976). See also *Annual Report: U.S. Atomic Energy Commission* (1974), and *Facts on Nuclear Proliferation* prepared by the Congressional Research Service, Library of Congress, for the Senate Committee on Government Operations, (U.S. Government Printing Office), December 1975.

[4] See, for example, the opening statements of Senators Ribicoff and Glenn, *Hearings, Senate Committee on Government Operations*, 94th congress, 1st session, April 24 and 30, and May 1, 1975 (U.S. Government Printing Office).

The basic reason for the adherence the Treaty has achieved has been the conviction that the danger of a world of nuclear powers is the greatest long run threat of the nuclear age; that the security of each nation would be strengthened far more by the commitment of each nonnuclear country of the world, including its local adversary, not to build nuclear weapons than by its own decision to maintain the nuclear option.[5] The second reason for adherence by many states was the prospect of obtaining special nuclear benefits as Treaty members. Article IV of the NPT specifically calls for the fullest exchange of information "among the parties" and for the provision of assistance by supplier states in a position to do so, particularly to Third World member states.

The familiar asymmetries of the NPT, in terms of the relative commitments of nuclear and nonnuclear weapons states, need no further amplification here; nor does the failure of the treaty to secure the accession of the dozen or so "near-nuclear threshold" states, such as Israel and Egypt, India and Pakistan, Argentina and Brazil—all mutual adversaries—and for somewhat different reasons, Spain, Switzerland, and South Africa. But the gradual accession of a number of additional threshold countries has been achieved in the past five years, and the asymmetry of commitment between weapons and nonweapons states was substantially reduced with the ratification of the NPT by the most advanced nonweapons industrial states in Europe in 1975, and by Japan.

Nevertheless, looking ahead in 1976, it seems unlikely that even the most imaginative initiatives coming from within the NPT regime can secure the accession of more than three or four of the critical threshold nations until the underlying military tensions behind their decision to maintain a nuclear weapons option are resolved.[6] In the meantime, the major NPT task in 1976 is to make sure that the threshold states conclude that they are better off inside the Treaty than outside it. At the moment, a formidable case can be made that the "nonjoiners" are better

---

[5] The present chapter, focusing on the contribution of controls over the transfer of nuclear technology to the nonproliferation regime, does not deal in detail with the most intractable obstacle to world-wide accession to the NPT—reflecting the underlying military tensions in many areas of the world—that of providing meaningful security assurances to nonweapons states in return for their commitment to reject the nuclear weapons option. The earlier assurances of the late 1960s by the U.S. and the Soviet Union, outside the treaty, to protect NPT members from the threat of nuclear blackmail or nuclear aggression, through prompt action taken in the U.N. Security Council, has lost its credibility because of the veto power of the People's Republic of China, a subsequent member of that body. For a new look at this problem and a provocative proposal to deal with it see Alfon Frye, *New York Times Magazine* (1976).

[6] See Chayes, "The Workings of Arms Control Agreement," *Harvard Law Review* (1972).

off and have little incentive to join up. This situation must be corrected by providing special nuclear power benefits to members of the NPT in return for their commitments not to build nuclear explosives. Furthermore, the principal suppliers must take far greater care in their commercial transactions to avoid providing comparable benefits to nonmember states.[7]

The importance of the nonweapons commitment of NPT members is often inadequately appreciated. Experience suggests that nuclear arms control agreements, while extraordinarily difficult to achieve, are not finally entered into with the objective of abrogating them. It should be remembered that India conducted its nuclear explosion *underground* because it had ratified the Limited Test Ban Treaty in 1963, and that it had chosen not to join the NPT because, among other reasons, the Treaty explicitly prohibited nuclear explosives, which it subsequently produced.

THE SAFEGUARDS SYSTEM

A major question about the NPT is whether the mandatory international inspection system, based on agreements negotiated between the IAEA and NPT members, is worth the paper they are written on. Specifically, critics question whether the IAEA is destined to be overwhelmed by the task of inspecting all the facilities of NPT nations and the individually safeguarded facilities of non-NPT states—not to mention the nuclear power facilities of the U.K. and the U.S., the two weapons states which volunteered to subject their own power facilities to international inspection.

Skeptics of IAEA safeguards have been numerous from the beginning, but their preoccupation with the possibilities that inspected nations may "beat the system" misses the central point of the assurance provided by safeguards. The purpose of the safeguards system has never been to make it impossible for nonnuclear weapons countries to separate the plutonium produced by reactors to build bombs. No foreseeable system can eliminate the possibility of diversion to military uses; it can only increase the likelihood of diversion being detected. And even as to detection, no system, without being impracticably intrusive, can approach certainty.

The central aim of the safeguards system is to make the likelihood of detection of "monkey business" sufficiently strong so that any country

[7] See Palfrey, "The Assurance of Safeguards" in Boskey and Willrich (eds.), *Nuclear Proliferation: Prospects for Control* (1970).

will think twice about trying to "beat the system." Detection of the secret diversion of fissionable material to build bombs would be very embarrassing for any nation, and that is precisely what the safeguards system aims to accentuate, as it applies to individually safeguarded facilities of non-NPT states and of all the facilities of NPT states. In the case of the latter, Article III of the Nonproliferation Treaty operates in conjunction with the solemn obligations undertaken by nuclear and nonnuclear weapons powers under Articles I and II, so as to provide more specific assurance that lawful, peaceful nuclear activities remain peaceful. The better the system becomes in the future, the longer a nonnuclear weapons state, party to the Treaty, will think about the consequences before secretly undertaking to divert fissionable material in inspected facilities. It can, under circumstances of supreme national interest, withdraw from the Treaty by the direct route of the Treaty's withdrawal clause. But formal action announcing withdrawal would produce international repercussions. Here again, such a nation would be aware of the perilous situation it might create by declaring its intention to develop nuclear weapons.

A SUPPLEMENT TO NPT—THE SECOND TRACK

In recent years it has become apparent that in addition to the NPT, a second track is needed to take care of the nonmembership of France (a current supplier), of the People's Republic of China (a potential supplier), and of the "threshold" countries developing nuclear power, such as Argentina and Brazil, South Africa and Spain, India and Pakistan, and Israel and Egypt. As a result of this situation, there are gaps in the nonweapons commitments and the safeguards coverage of the NPT regime. While specific equipment and materials acquired by non-NPT states from member states are subject to IAEA safeguards, those that are indigenously constructed, or received from a non-NPT supplier state, may not be safeguarded.

In the past two years, in an effort to reduce these gaps and to cope with the competitive suppliers markets, the U.S. has undertaken to develop a common front among the principal supplier nations, including France—a non-NPT state—on the safeguards and restrictions to be applied to all commercial transactions with all Third World countries. It is publicly known that there have been conferences among the principal suppliers, but it was not until February and March 1976 that government spokesmen provided the first authoritative reports of the results of these discussions. On February 23, 1976, Dr. Iklé, Director of the Arms Control and Disarmament Agency (ACDA) stated to the Disarmament Sub-committee of the Senate Foreign Relations Committee that

> The United States over the years has sought to work with other countries to insure that civil nuclear exports would be used only for peaceful purposes. We have recently had a number of bilateral and multilateral discussions with nuclear exporters to develop common rules on safeguards and export controls. As a result, the United States together with other exporters has decided to apply certain principles to our future nuclear exports. Most of these are consistent with current U.S. practice; some are new. All are designed to inhibit the spread of nuclear weapons while permitting nuclear exports of equipment to meet the world's growing energy needs. These principles include the following:
>
> [1]—The requirements that recipients must apply international (IAEA) safeguards on all nuclear imports.
> [2]—The requirement that the importer give assurances not to use these imports to make nuclear explosives for any purpose—whether called "peaceful" or not.
> [3]—The requirement that the importer have adequate physical security for these nuclear facilities and materials to prevent theft and sabotage.
> [4]—The requirement for assurances that the importers will demand the same conditions on any re-transfer of these materials or types of equipment to third countries.[8]

Application of the first and fourth principles should go far toward removing safeguards considerations from the competition among the major suppliers in the sale of nuclear equipment and materials. In the course of Dr. Iklé's statement, as amplified by Dr. Kissinger's testimony of March 9,[9] it was made clear that the commonly imposed safeguards would be extended in terms of their application to the "technology" involved in the export of facilities. The net effect would be to extend the requirements of safeguards not only to retransfers of materials or equipment to other countries, but also to any replication of that technology in subsequent plants built by the receiving country, or in plants into which "weapons usable material" derived from an import may pass.

The second principle was designed to remove any questions about the receiving country using the imports to build nuclear explosives and calling them "peaceful," as India undertook to do with its explosion in 1974. Except for its intended use, a peaceful nuclear explosive (PNE)

---

8 Hearings of the Disarmament Subcommittee, Senate Committee on Foreign Relations (yet to be published).

9 Hearings of the Senate Committee on Government Operations (yet to be published).

is indistinguishable from a nuclear weapon. The IAEA has already specifically stated that in its safeguards arrangements to ensure that the materials, equipment, and information are not used to further "any military purpose," peaceful nuclear explosions are included in this category. Finally, the NPT in Article II contains a specific commitment by nonweapons states not to manufacture or acquire nuclear weapons, "including peaceful nuclear explosives."

The third principle concerned that aspect of safeguards relating to the physical security of the plants, with particular attention to protection against nongovernmental terrorist groups. This has become a matter subject to bilateral negotiation between the supplier and receiving country, and for interesting reasons of national sovereignty relating to basic national security concerns has not been made a part of the IAEA safeguards process. The U.S. has recently taken the lead in requiring that the supplier be satisfied with the physical security measures of the receiving country. All countries presumably have a common interest in establishing adequate physical security. In view of its importance, the IAEA may take the initiative in holding a multilateral convention leading to the establishment of agreed physical security standards. Enlarging on these basic principles, Dr. Iklé went on to say:

> Together with other leading exporters of nuclear technology, we are also committed to follow-up efforts along three lines.
> 1. To promote international cooperation in exchanging information on physical security, on measures of protection of nuclear material in transit, and on measures for recovery of stolen nuclear material and equipment.
> 2. To improve the effectiveness of IAEA safeguards through special efforts that support that organization, and
> 3. To encourage the designers and makers of sensitive equipment to construct it in a way that will aid safeguards.

Dr. Iklé's statement made no reference to any requirement that suppliers condition the transfer of fuel and equipment to non-NPT countries on their acceptance of IAEA safeguards on all their facilities, indigenous or acquired (the so-called "poor man's NPT," accepting the full range of safeguards, but without a formal nonweapons commitment under the NPT). It is understood that the IAEA may consider measures to implement such an approach, if adopted by supplier countries, but the course of discussions among the suppliers on this issue has not been disclosed.

While progress in achieving a common front among the suppliers on this issue is unknown, there have been reports of Canadian and British support for such an approach. The obvious question is the position of

France—a non-NPT member—on this issue. The public indications are that France has long supported IAEA safeguards, and according to the recent U.S. accounts of the results of the suppliers conferences, is now prepared, along with the other suppliers, to strengthen safeguards and extend their reach, but without reference to the NPT. How France might view a "poor man's NPT" is not known, but a requirement of IAEA safeguards that extends to the indigenous facilities of nonweapons states, including those not involved in a specific transaction, might be regarded by France as the same kind of discriminatory treatment that led it not to join the NPT in the first place.

Even if France were to accept such an approach, however, there is a very real question whether a requirement of mandatory safeguards, without a nonweapons commitment, would not "debase the currency" of the NPT, which requires such a commitment. The "poor man's NPT" could well provide "an easy way out" for a number of countries seriously considering joining the NPT, who are prepared to accept mandatory safeguards, but who are still hesitant about taking the formal step of rejecting the nuclear weapons option for the indefinite future.

CAN EXPORTS OF REPROCESSING TECHNOLOGY BE JUSTIFIED?

Dr. Iklé dealt circumspectly with one major issue—the question of common supplier export policies on the sale of national fuel enrichment and reprocessing plants, for example, the FRG-Brazil agreement, the cancelled French-South Korean sale of a reprocessing plant, and the current French-Pakistan negotiations for a reprocessing plant. On this subject, Dr. Iklé merely stated that the U.S. intended to use "restraint" in the export of "sensitive" technology such as fuel enrichment and reprocessing, particularly in cases where such exports could add to the risk of proliferation; and to require U.S. consent over their reexport to any third country. He stated these were *minimum* standards—possibly reflecting the current extent of a common suppliers front—adding that the U.S. was prepared to be "more stringent" when appropriate.

It has been the general policy of the United States not to authorize the export of either technology. The difficulty is that two other suppliers, France and West Germany, have pursued a different course, over the objections of the United States.

The problem of immediate concern is that a market has developed among the "threshold" countries for the acquisition of reprocessing plants for the avowed nuclear power purpose of "learning the art," preparatory to future recycling of the plutonium in current reactors, with prospective savings as a result in the uranium resources and the enriched uranium fuel required. Because of this interest, however, a direct conflict between commercial and nonproliferation considerations has been posed for suppliers. There has been a developing tendency on the part of some suppliers and receivers to tie the acquisition of a small reprocessing plant with an agreement to purchase nuclear power reactors from the supplier country. While the profits from the sales of such reprocessing plants are comparatively very small, the profits from the sale of power reactors, now and in the future, may amount to billions of dollars.

Except for one consideration, noted hereafter, the nonproliferation argument against the sale of reprocessing plants to threshold countries is very strong. In terms of any nuclear power objective, such plants are totally uneconomical and technologically premature for countries with only a few light-water reactors. Economic studies have shown that even in countries with major nuclear power programs, reprocessing plants have no economic justification unless they are very large. And even if they are large, recent U.S. experience has indicated that unsolved problems of escalating cost, technology, safety, safeguards, and environmental factors affecting the so-called "back-end" of the nuclear fuel cycle suggest that it may make more economic sense simply to store the spent fuel in recoverable form before embarking on a course of large-scale reprocessing to separate the plutonium for recycling. Reflecting these realities, authorization for full-scale plutonium recycling has been withheld by the U.S. Nuclear Regulatory Commission, pending further study.

Therefore, the acquisition of reprocessing plants by developing countries constitutes a highly provocative act in nonproliferation terms because it represents a decision to accumulate plutonium that is immediately usable only for the construction of nuclear explosives. Dr. Iklé labeled such a step in just that way when he attacked the proposed French sale of a reprocessing plant to Pakistan as an act of proliferation on the part of Pakistan, in response to the Indian nuclear explosion.

There is one nonproliferation argument *in favor* of the sale of reprocessing plants, and it has been emphasized by the Germans in support of the Brazilian agreement. The construction of a small reprocessing plant is not beyond the capabilities of a threshold country, and can be accomplished indigenously, as India demonstrated. Therefore, for a non-NPT country whose facilities are inadequately safeguarded, the most important objective for NPT suppliers is to make sure that if such plants are likely to be built anyway, they should be subject to the most strenuous safeguards possible. The alternative could well be reprocessing plants built indigenously, or with help from non-NPT parties, with no safeguards of *any* kind.

The German position is not easily refuted because reprocessing plants by a determined country with modest industrial capabilities can be constructed without outside help, although plutonium separation is by no means a simple process. The principal argument against this German position is that if suppliers refused to sell national reprocessing plants, it might take many years longer for a Third World country to construct such a plant on its own, with no outside help (assuming no assistance from a non-NPT state such as India or China). Furthermore, the technical difficulties experienced by Argentina in constructing reprocessing plants on its own, and the technical training course that Indian scientists received from the U.S., suggest that while the literature on reprocessing is widely available, the actual construction and the safe and efficient operation of a plutonium separation plant may be more difficult than is commonly supposed.

The German approach is partly consistent with the pursuit of a "poor man's NPT" in the sense that it requires all the facilities received from West Germany, including subsequent facilities based on the technology transferred, to be placed under IAEA safeguards. But it falls short of that pursuit because it does not include a commitment by the receiving country to place *all* its facilities under safeguards, wherever received and however constructed.

The German position, moreover, is inconsistent with the provisions of Article IV of the NPT, and the objectives of the NPT in general, because it provides special benefits to a non-NPT member (a complete fuel cycle), thus effectively eliminating *any* incentive for Brazil to join the NPT, and to make a formal nonweapons commitment. Even if other suppliers refuse to deal further with Brazil—a most unlikely prospect since U.S. firms, for example, are actively interested in selling reactors to Brazil—Brazil can count on Germany to meet its future nuclear power needs. Should the German "jet nozzle" enrichment plant prove to be unsuccessful or inadequate to provide Brazil with enriched uranium by the 1980s, it is likely that the Germans will then be capable of providing enriched fuel through its gas centrifuge plant.

Multinational Fuel Centers (MNCs)—the Affirmative Case

It is in the context of disagreement among supplier nations that the United States has recently insisted on retaining in its reactor sales the right to approve where reprocessing may take place. In addition, the

United States, through Dr. Kissinger's speech to the General Assembly in the fall of 1975 (and amplified in his congressional testimony of March 1976) has proposed the establishment of multinational fuel centers (MNCs) as an alternative to the sale of reprocessing plants. An IAEA feasibility study of MNCs is currently underway, strongly supported by the United States.

The central philosophy of the MNC concept is that economic, technological, and nonproliferation considerations coincide. It provides a major reinforcement of IAEA safeguards, when accompanied by supplier involvement in the facilities of threshold countries, by establishing nuclear fuel centers to serve the needs of more than one nation. Such centers could concentrate first on providing centralized and technologically sophisticated facilities for storing spent fuel. Later, as the nuclear power market develops for the countries involved, and if the technology and economics of plutonium recycle are demonstrated, and genuinely effective international controls over the separated plutonium are established, the fuel center could be expanded to include large-scale reprocessing plants, mixed oxide fuel fabrication, and long-term waste storage facilities, in colocated centers, hopefully under IAEA auspices, and possibly possessing extraterritorial status.

The multinational approach is highly flexible, in terms of participants and location. It need not be regional. For example, the center might involve a major supplier country not only in the role of providing the design and construction of an exported facility but also as a participant in a joint venture to obtain reprocessing services to meet its own needs. As discussed hereafter, the multinational center might be located in the territory of the supplier country. It may prove easier to achieve agreement on a fuel center among nations in different regions than among suspicious neighbors. Alternatively, suspicious neighbors may be more prepared to join in such an enterprise, to monitor each other, with the presence of a participant from another region.

Moreover, nuclear precedents have already been provided for multinational ventures in the uranium enrichment field by the French lead gaseous diffusion plant (Eurodif.), involving participation of Italy, Spain, and Iran, and by the consortium of England, Germany, and the Netherlands in developing enrichment plants using gaseous centrifuge (Urenco), and in the reprocessing field by the earlier venture of "Eurochemic."

The nonproliferation advantages of multinational centers are obvious in complicating takeover by nationalization and subsequent diversion of the separated plutonium by any individual country. Safeguards against diversion of nuclear materials could be more effective in an MNC than in a national facility because each of the MNC participants could serve as a monitor to guard against diversion of materials by other participants. Furthermore, the costs of implementing IAEA safeguards and providing physical security against theft and sabotage would be lower for a single large MNC than for several national facilities. Greater flexibility in the choice of location also could contribute to the MNC's physical security advantage. Participation in such a center would not significantly simplify the task of any one receiving country thereafter designing an unsafeguarded national plant of its own.

Among other important nonproliferation advantages of MNCs are (1) that they provide for the potential evolution of the role of the IAEA, both under the NPT and also under the large and broadly conceived terms of its original charter; and (2) that the MNC approach comes close to reconciling the differences in assumptions reflected in Article IV of the NPT, calling for the fullest possible exchange of technology among the member states, and the thrust of the current U.S. approach, calling for "restraint" when exchange of technology is regarded as adding to the risk of proliferation.

It would be difficult for the IAEA, under its own charter, and in implementing the NPT, through negotiated safeguards agreements with the member states, to act as the instrument of *restricting* the peaceful development of nuclear power. It would not be difficult, however, for the agency to participate in a program of assistance in the appropriately staged development of nuclear power that also simplifies and reinforces its own capabilities to safeguard the facilities and materials involved.

The importance attached by the U.S. to international recommendations and to international feasibility studies suggests that the U.S., having initiated the multinational concept, as an accompaniment (or an alternative) to flat opposition to sales of national reprocessing plants, believes that its evaluation and possible implementation should come from the IAEA, representing advanced and developing countries, suppliers and purchasers. Thus, Secretary Kissinger's MNC proposal to the U.N. should be seen as the launching of a promising evolutionary concept, whose future should depend not on pressures from the U.S., nor on pressures from the principal suppliers (if they should support this approach), but on world-wide recognition that MNCs provide a functional, practicable, and flexible route to assist in the appropriately staged development of nuclear power to meet the world's energy needs, and at the same time to provide an important reinforcement of the nonproliferation regime.

One obvious question is whether the concept of MNCs could not only bridge the gap in assumptions behind the NPT and the supplier conferences, but also the gap in intentions and practices among some of the principal suppliers. There is a suggestion in Dr. Iklé's recent testimony that the position of suppliers on this issue may be influenced by the outcome and reception of the IAEA feasibility study, scheduled for completion in 1977.

Multinational Fuel Centers—the Negative Case

Many questions, however, have been raised about the MNC concept. Is it "gimmickry" and a transparent device to seek to impose "cartel-like" controls on threshold countries? Is its basic aim to block threshold countries' development of the full potential of nuclear power including plutonium recycle and breeders, which the advanced countries are actively pursuing?

The proposal has been regarded by many supplier and receiving countries simply as a complicated "fall back" position on the part of the U.S. from that of its flat opposition to any sales of national reprocessing plants by the principal suppliers to threshold countries, NPT members or not. Since the U.S. could not prevent such sales by others, thus placing itself at a competitive disadvantage in reactor competition, it was regarded as trying to "buy time" by persuading suppliers to defer such sales until they constituted part of a multinational enterprise. Inevitably, the proposal revives memories of the ill-fated Multilateral Nuclear Force proposal for NATO in the 1960s. What success the U.S. has had in developing a common front among suppliers on this subject was not disclosed at the recent hearings. There are very real questions as to the acceptability of such a concept either among the principal suppliers with an immediate market for national plants, or for the receiving countries, who might dislike involvement in a regional venture, or in a venture dominated by an advanced country participating.

The most fundamental question is whether a multinational center, if successful, would *attract* Third World interest in reprocessing, at a time when our principal interest is in deferring their premature involvement in the "back-end" of the fuel cycle. Dr. Iklé was quite frank about the problem in saying that U.S. interest in multinational approaches should not be misunderstood. "We do not wish to promote the reprocessing of plutonium. On the contrary, our hope in all these efforts is to investigate practical, economic alternatives to national reprocessing and thereby reduce the growing dangers of nuclear proliferation."

CENTRALIZED REPROCESSING FACILITIES

In view of this amplification of U.S. objectives, and of the skeptical initial response to the MNC concept, the central point emerges that MNCs should not be regarded as *the* U.S. solution to the reprocessing problem. The tail should not wag the dog. The underlying objective of the U.S. is to convince Third World countries that at least for the next two decades, they can meet their nuclear power needs by concentrating on the well established "front-end" of the nuclear fuel cycle, and that reprocessing and plutonium recycle are highly premature and totally unnecessary. They can fully meet their prospective needs by the purchase of the proven light-water reactors, fueled by an assured supply of low enriched uranium, supplied at reasonable cost, with appropriate arrangements and assistance provided for the storage of spent fuel.

There are a number of alternative approaches to be considered to cope with the disparities in the time scale of demand for reprocessing between advanced and developing countries. One potentially attractive alternative could be the centralization of reprocessing and mixed oxide fuel fabrication facilities in the form of a multinational or highly coordinated enterprise, located in the territory of the supplier countries and designed to meet the immediate needs of those advanced countries who have (wisely or unwisely) already embarked on a course of plutonium recycle and of accelerated breeder development. The objective would be to establish a capability of providing centralized reprocessing and fuel fabrication services facilities for the world (as is the case with uranium enrichment services), and to obviate the need for uneconomic and premature plants in various regions elsewhere.

The enterprise would be open to Third World Investment participation which would provide those countries with the assurance of reprocessing and fuel fabrication services at a later date, when the development of their nuclear programs warrant it. At that time, the distribution of the mixed oxide fuel elements, following reprocessing, might be funneled through the IAEA, providing assured transportation safeguards. A further advantage of this approach is to provide greater protection against easy access to the separated plutonium, prior to its fabrication into mixed oxide fuel elements, which would thereafter be in a form far less accessible in terms of diversion for nuclear explosive purposes.

PROSPECTS FOR AN AGREEMENT AMONG SUPPLIERS—A NUCLEAR CARTEL?

One "nonsecret" characteristic of the suppliers conferences has been their secrecy, even as to the identification of the states participating in them, and the quite obvious disinclination, reflected in Dr. Iklé's testimony, to classify their outcome as a formal agreement. The fact that Dr. Iklé made no reference to the relationship of the conferences to the NPT is evidence of the delicacy of the problem of dealing with nuclear exporters and importers, not all of whom are members of the NPT, and of restrictions on "sensitive" technology, not all of which would be consistent with the "fullest possible exchange" of technology among members of the Treaty, as prescribed in Article IV.

Dr. Iklé made it clear that while the U.S. had no particular brief for secrecy it was genuinely interested in making headway in developing a common front among suppliers; and if particular countries had their own internal reasons for secrecy, the U.S. would respect them, as a prerequisite to the existence and continuation of these consultations. A final reason for the secrecy of the meetings, and the ambiguities of the outcome, could reflect a desire on the part of the participating states to avoid the appearance of establishing a "nuclear cartel" by the advanced nuclear power states.

But would a nuclear cartel be so undesirable? In testimony before the Senate Committee on Government Operations, Dr. Steven Baker, after pointing out that a nuclear monopoly is no longer possible, if it ever was, and that a free market is not a realistic option because nuclear power is regarded as too important by the governments to be left to the operation of market forces, recommended the formal establishment of a "nuclear consortium" of the principal suppliers to provide enrichment and reprocessing services on a commercial basis (with emphasis on enrichment), and with competition among the consortium limited to *reactor* exports. He claimed that the London Supplier Conferences have already been labeled an incipient cartel, thus leaving nuclear suppliers with all the practical liabilities of a cartel but with none of the political benefits.

A nuclear fuel cartel, in Dr. Baker's view, would be inhibited from abusing its power because if it manipulates its terms of supply of enriched uranium, importers would turn to natural uranium reactors such as Canada's "Candu" reactors, or form their own enrichment and reprocessing ventures, national or multinational. Thus, Dr. Baker distinguishes this nuclear "technology based" cartel from a "resource based" cartel, like OPEC. He concludes that if the nuclear cartel succeeded in offering long-term supplies at low prices, there would be little incentive to develop national enrichment facilities. Moreover with adequate enriched uranium there would be little incentive for reprocessing and plutonium separation to provide supplementary fuel.

However refreshing in their candor, Dr. Baker's views are unlikely to be adopted by the principal supplier nations, who seem peculiarly indisposed at this time to take on the Third World in this fashion. For Third World nations, having been driven to develop nuclear power because of the Middle East stranglehold on oil, buttressed by the OPEC cartel, to discover suddenly that they are faced with a nuclear cartel as well (prohibiting exports of enrichment and reprocessing plants, controlling supplies of enriched uranium to fuel their reactors, and prohibiting their acquisition of national reprocessing plants designed to reduce their dependence on uranium ores and enriched uranium), the result could be a political explosion and an increasing polarization between developed and developing nations that would not be pleasant to contemplate. It might also seriously undermine whatever Third World support for the NPT that continues to exist.

THE CENTRALIZATION OF URANIUM ENRICHMENT SUPPLIES

This is not to say, however, that something approaching a functional centralization of supply capabilities may not be in the offing in the field of uranium enrichment in the course of the next decade. The basic concept, however, would be very different from a cartel because Third World countries would be participants in multinational enrichment plants to be assured of long-term fuel supplies even if their participation is variable and sometimes limited to financial investment.

Such a centralization of enriched uranium supplies would probably be easier to establish than a centralized reprocessing and fuel fabrication enterprise in view of the existing dependence of Third World countries on enriched fuel from the suppliers, the substantial centralization that already exists, and the expense of enrichment plant construction for developing countries. Furthermore, the multinational "Eurodif" gaseous diffusion plant, that includes the investment participation of the Third World country of Iran, provides a precedent for the extension of such a concept in the form of a larger coordinated suppliers venture.

The German-Brazilian agreement which included not only the sale of a reactor and a reprocessing plant, but an enrichment plant as well, using the German developed "jet nozzle" process, is not in the interests of the nonproliferation regime for the reasons already discussed—because it provides a complete nuclear fuel cycle for a non-NPT member. However, it should be recognized that the agreement, as it relates to the provision of an enriched uranium route to nuclear explosive materials, poses a less imminent threat to proliferation because of the huge expense in power and money involved for the receiving country to acquire an independent capability. So long as light-water reactors are predominant, using only slightly enriched uranium, almost all of the fuel for the next decade will be purchased from the U.S., produced in its gaseous diffusion plants, and gradually supplemented by West Europe in its uranium enrichment plants under construction, and to some extent by the Soviet Union and potentially by Japan and eventually by other suppliers.

As long as threshold nuclear countries are assured of adequate long-range fuel supplies, the technologically difficult assignment of building an indigenous enrichment facility, and the vast expense in money and power of acquiring and operating such a plant, provides a substantial disincentive to the early development of a significant market in this field. The only commercial justification for the Brazilian acquisition of the German "jet nozzle" enrichment plant is that it happens to make use of a large untapped Brazilian potential of hydroelectric power.

The question for the major suppliers to face, however, is whether transactions to construct *any* enrichment plants outside their borders should be encouraged, however tightly safeguarded. A strong case can be made (at least for the present) in view of the expense, technology, classification, and nonproliferation significance of such plants, that multinational plants to supply enriched uranium should be constructed *within* the territories of the supplier countries. They should include investment participation by threshold countries to assure themselves of long-range fuel supplies, without the uneconomic expense of having to construct such large plants in their own territory for a limited number of reactors.

The situation may change as nuclear power programs enlarge, and particularly with the development of less expensive alternatives to gaseous diffusion, such as centrifuge and ultimately laser technology. Every effort should be made to use the leverage of the advanced suppliers in developing economically competitive alternatives to gaseous diffusion, to control the transfer of this technology, and safeguard its use. While thought has been given to the possible inclusion of enrichment facilities in MNCs, there is a strong case against it.

Indeed, for the immediate future it would be better to concentrate on multinational fuel centers devoted *first* to the storage of spent fuel, and to steer clear of the concept of complete nuclear fuel cycles, including enrichment plants, reprocessing, and plutonium recycle in various regions of the world. The notion that the U.S. unwittingly conveyed to the world by its own activities, that plutonium recycle and ultimately breeder reactors are the inevitable wave of the future, did a profound disservice to the cause of nonproliferation by providing an impetus for countries to acquire reprocessing plants as a *reputable* nuclear power objective, while also providing them with a nuclear weapons option. We would compound our error if we conveyed by our actions the notion that nations should have regional enrichment facilities as well (if we have not already done so by failing to guarantee long-term fuel supplies).

On the other hand, as discussed hereafter, we must do whatever we can to recover some of the leverage we lost in failing to provide first, for ourselves, and second, for the world, an adequate low-enriched uranium supply capability. Since Western Europe and Japan are now determined to assure their own supply for the longer haul, our obvious choice is to work with them to control world-wide nuclear power development through the provision of adequate fuel supplies. We should approach the question in conjunction with Western Europe and Japan, and probably the Soviet Union. The undertaking should be a joint one and should provide at the very least for the investment participation of Third World countries. Such an initiative would be likely to have a generally favorable reception among the principal suppliers as a venture in joint planning, since they need the fuel for themselves, and since it offers the prospect of a profitable overseas market, as well as being central to a strengthened nonproliferation regime. For Third World countries it would provide the most practical route to an assured supply of enriched uranium.

Unilateral Action by the United States

Any discussion of what is possible by way of unilateral action must start from the premise that there is no way of stopping countries with modest industrial capacities from developing nuclear weapons if they regard it in their supreme national interest to do so; and over thirty years it has become progressively easier to acquire nuclear explosive materials.

But granting that fact, is there some action which the U.S. can take to make it as hard as possible? In April 1975, Senator Ribicoff asserted that the nuclear export policies and practices of this nation "are not only dangerous; they are scandalous." More recently he said: "Why don't we go to the United Nations and world opinion on this horrible question? Isn't it time to go over the heads of the leaders to the people of France and Germany—I see no reason for the U.S. to continue to be so timid." Then, proceeding to specifics, he proposed that the U.S. join with the Soviet Union in making use of their monopoly of enriched uranium during the next seven or eight years, to threaten to cut off enriched uranium fuel supplies to France and West Germany unless they stop the sale of reprocessing plants abroad. (In the absence of Soviet agreement, Senator Ribicoff recommended unilateral action by the U.S.)

Secretary Kissinger replied that such an initiative, with Soviet concurrence, would amount to U.S. blackmail against NATO allies, and would have the "gravest foreign policy consequences." He said further, "We are prepared to cooperate with the Soviet Union, together with our allies. The only point is, we are not prepared to cooperate with the Soviet Union against our allies." Secretary Kissinger then made it clear that the U.S. would continue to pursue the issue of sales of national reprocessing plants at subsequent suppliers meetings. Further, he commended the Soviet Union for exerting a "positive moral and political influence" at these meetings.

A final problem with Senator Ribicoff's proposal is that it transforms the nonproliferation regime into the very vehicle that supporters of the NPT have been denying that it was—nothing more than a raw power play by the nuclear weapons states to impose their will on the non-weapons states and reap all the profits. Having just strengthened the case that it is not such a vehicle by securing the ratification of the most advanced nonweapons states, to suddenly apply a dubiously successful form of nuclear power blackmail would make a mockery of the whole enterprise. The subtle exercise of diplomatic leverage is difficult enough; the use of outright threats is quite another matter.

There was general recognition in 1976 that the world-wide commercialization of nuclear power had proceeded to the point where unilateral actions by the U.S. to withhold the transfer of its own technology would result primarily in the business flowing to the competitors in view of their independent commercial capabilities. Thus, any unilateral proposal for the U.S. to require a form of a "poor man's NPT" for those nations that have not made a nonweapons commitment by NPT ratification (by requiring their commitment to accept safeguards on all their facilities, indigenous or otherwise, in return for U.S. exported fuel or equipment) would do little more than lose business for the U.S. government and its nuclear industry, unless the other principal suppliers adopted the same policy—which is currently unlikely.

Even in the area where the U.S. still retains a temporary residual monopoly, in the provision of enriched uranium fuel, experience from

1970-74 provides a cautionary tale in terms of the further exercise of unilateral leverage by the U.S. On the one hand, there was some evidence of arm twisting on our part, based on our monopoly of enriched uranium, to accelerate NPT ratification by the Euratom countries and by Japan. This was not well received, particularly when we offered to provide reactors and enriched uranium to Israel and Egypt—both notorious non-NPT members. Yet, at the same time, the U.S. was engaging in an extraordinarily vacillating performance with regard to the need for, and the form of, provision of an assured supply of enriched uranium for itself, and others.

In fact, one of the strongest indictments of recent U.S. nonproliferation strategy has been its failure to recognize and to take account of the international implications of its actions in its domestic nuclear power planning. The most prominent example has been in the uranium enrichment field. Enmeshed in the domestic politics of the "privatization" of the entire nuclear fuel cycle (already demonstrated to be anything but a conventional opportunity for private investment) the AEC, and its successor ERDA, from 1969 to 1975, were thinking primarily in terms of meeting domestic uranium enrichment needs, as reduced by the prospect of fuel savings through extensive plutonium recycle; and of the potential role of competitive private enterprise in making the charges for the "separative work" more nearly reflect the actual costs of uranium enrichment. The result was an interval of six years during which the government waited around for industry to complete its studies on the possible "private takeover" of existing government gaseous diffusion plants, and thereafter, on the private construction of a fourth diffusion plant.

Whatever the conceptual merits, a critical delay ensued, with incalculable damage to U.S. nonproliferation strategy, in the launching of a major program of upgrading existing plants and the construction of a fourth gaseous diffusion plant in order to provide an assured long-term supply of enriched fuel for the domestic and foreign market by the 1980s. During the same period, the AEC abruptly raised the prices for "separative work" on several occasions, without consulting the State Department or its overseas customers, and it established new and demanding financial conditions for all further "toll enrichment" arrangements with foreign nations.

The principal outcome of this interval of delay and, heavy handedness on our part was to convince the Europeans of the unreliability of our highly touted assurances of a long-range supply of enriched uranium to fuel the reactors we, and they, were buying, selling, and using. The result could have been foreseen—an acceleration in the efforts of the advanced countries to provide independent sources of enriched uranium, and of plutonium recycle, and an exacerbation of relationships among the U.S. and the other principal suppliers, leading to a determination on the part of European nuclear industries to exploit the lucrative competitive nuclear power market opening up, if necessary, at the expense of nonproliferation priorities.

This experience would strongly indicate that the continued pursuit of the recent collaborative efforts of 1975-76 among the *governments* of the principal suppliers, to strengthen nonproliferation priorities to be exercised in the course of commercial competition, along the lines already discussed, is a much more promising approach than implicit threats based on our residual enriched uranium leverage to make the other suppliers behave as we would like them to—particularly since it makes the case for reprocessing and recycle, pending development of their own enrichment capabilities, that much more compelling. Apart from the specific results of the suppliers conferences, there have been some indications in the foreign press that top level governmental attention is finally being given to seek ways of making commercial considerations the instruments, not the masters, of nonproliferation objectives.

It is difficult to determine how much credit for this change of attitude should be given to the public outcry in the U.S. Senate and in the press against the West German-Brazilian agreement, after it was signed, and against the proposed French sale of a reprocessing plant to South Korea, which was cancelled, and the similar French sale to Pakistan, which may be too late to stop. On the one hand, public criticism may have led top officials of the U.S. government to adopt a harder line on the reprocessing issue (which they deny), and may have led to "second thoughts" about such sales on the part of France and Germany (which they will not admit). On the other hand, the evident French and West German anger at U.S. proposals to threaten a cut off of nuclear fuel, and at comments labeling them as "the nuclear proliferators" could well backfire, and lead to intransigence at future suppliers' conferences.

It is this writer's view that if the U.S. message of top level opposition to the sale of enrichment and reprocessing plants was inadequately conveyed a year ago, the message has certainly gotten through by now. The time has therefore come for the U.S. to lower its voice in public and to return to private diplomacy with concentration on imaginative initiatives directly aimed at the central problem of proliferation—the establishment of agreed controls over the transfer of technology that limit the access to nuclear explosive capabilities provided to Third World nations, while also providing assistance in meeting their nuclear power needs. This would not mean any let up in U.S. opposition to the sale of national reprocessing plants, nor to the possible exchange of German technology with Iran, for instance. What does need exploration are ways to enable the West European countries to compete effectively in the sale of facilities and equipment to Third World countries without tearing apart the nonproliferation fabric in the process, by using "sweeteners" that involve the transfer of nuclear explosive technology, however heavily safeguarded. If everyone followed the same ground rules of restraint on this point, it would not seem to be necessary to engage in a cartel-like approach to the sharing of the Third World market for nuclear power reactors by the principal suppliers.

U.S. DEFERRAL OF PLUTONIUM RECYCLE

There remains one initiative available to the U.S. which is not high handed. This initiative, while the nonproliferation priorities are sharply put into focus, is still in keeping with the organic development of the nonproliferation regime over the past twenty years. In fact, this initiative is consonant with the story of successful arms control ventures generally, which have depended, first, on the development and identification of a sufficient common interest among the participants, which *then* makes the arms control calculus of the advantages of reaching an agreement the deciding factor.

The proposal is that the U.S. government should decide to defer, for a specified period of years, the domestic licensing of plutonium by the nuclear industry for use as a secondary fuel in the current light-water reactors. While the government would proceed with a demonstration program of plutonium recycle for commercial use in current reactors, as a hedge against unforeseen escalation in the costs of uranium and of enriched uranium fuel, the primary emphasis of the government program would be on the use of plutonium as the primary fuel for the breeder reactors, when their commercial feasibility is demonstrated. Concomitantly, the U.S. would embark on a major program of government support for intensive exploration and extraction of uranium ores, and provide a major enlargement of its uranium enrichment capabilities—in this writer's opinion, by government construction, rather than by federal guarantees for private construction—not only to meet domestic needs, but to lay the foundation for the collaborative effort among the principal suppliers to meet the world's needs for enriched uranium.

As noted in previous chapters, it is a close question whether the reprocessing and recycling of plutonium, in the form of mixed oxides, to fuel the current light-water reactors is a preferable commercial alternative to the continued reliance on low-enriched uranium—*assuming* a major increase in uranium exploration and extraction and in enrichment capabilities.

Commissioner Gilinsky of the Nuclear Regulatory Commission in testimony before the Senate Committee on Government Operations made a number of important points in this regard. He testified at a time when his own commission had yet to decide whether to license the use of recycled plutonium for wide-scale use in current reactors in the U.S., and he properly focused on U.S. export controls in the interim period. But his awareness of the international implications of plutonium recycle was prominent:

> While plutonium recycle remains an open question in this country, and while even a favorable decision may have little relevance for nations with smaller nuclear programs, the interest in this alternative on the part of advanced nuclear nations has conveyed to other countries of the world, both large and small, the notion that plutonium recycle is an essential feature of an economically viable nuclear program. The prospects for effective coordinated international action on secure disposition of plutonium will therefore be strongly influenced by the apparent value of plutonium for recycle. *It is seldom pointed out that what is actually at stake is a relatively small fraction of nuclear fuel cycle costs, and an even smaller fraction of power costs—perhaps three percent in all.* [emphasis supplied] There are some preliminary indications, in addition, that the penalty for as much as a several-year delay in plutonium recycle may be relatively small. This would imply that immobilization of international plutonium stockpiles for that much time might not affect nuclear power economics significantly. In other words, it may be possible to enhance international security without injury to vital national economic interests.

Suppose, therefore, that the U.S. decided that nonproliferation considerations tipped the scales, because the economics and the technological and environmental considerations associated with immediate plutonium recycle for current reactors have already made it a marginal commercial venture (including the fact that the "plutonium cycle" is a major source of opposition from the domestic nuclear power critics). The U.S. in announcing the decision to defer reprocessing would make it clear that while the decision was justified on the domestic commercial merits of the case, and the unresolved environmental and safeguards issues, the decisive factor was its pursuit of "the paramount objective of assuring the common defense and security" in terms of highlighting the danger of reprocessing plants involving plutonium separation, and of fortifying the U.S. position against the sale of national reprocessing plants to Third World countries. In terms of the U.S. domestic program, there is no reason to believe that government assistance in providing for spent fuel storage, in retrievable form for a number of years, is beyond the technological capabilities of the U.S. to achieve, once it makes up its mind. Canada has been storing spent fuel from its natural uranium reactors, with steadily improving capabilities.

One important contribution of this decision would be to strengthen the NPT regime by reducing its familiar asymmetries between the undertakings and commitments of nuclear weapons states and of nonnuclear weapons states. To a developing country which had ratified the NPT, it could appear downright insulting, having made the commitment not to build weapons, and to accept mandatory safeguards on all its nuclear power facilities in the expectation of forthcoming nuclear power benefits under Article IV, to be told *then* that it cannot acquire national reprocessing plants, while all the advanced countries are actively engaged in such reprocessing. The economic argument that countries with only modest nuclear power programs are not ready for reprocessing and plutonium recycle, even though entirely valid, could have a hollow ring to it in this context. However, if the leading nuclear power of the world publicly decided to defer reprocessing, the impact could be immeasurable.

Such an initiative would not be inconsistent with the exploration of multinational fuel centers, but it would indicate that the first step should probably be the establishment of spent fuel storage centers to meet the early needs of threshold countries, in the form of international depositories set up under the auspices of the IAEA and conceivably operated by the agency as an IAEA center, possessing extraterritorial status. Time would thereby be provided for the IAEA, with the support of member nations, to give the most intensive consideration to the best ways of providing a genuinely effective system of international controls over the back-end of the nuclear fuel cycle, at a later date, when there is a justifiable demand for its exploitation by Third World countries.

Furthermore, such an initiative would not prevent the U.S. from participation in the exploration of a centralized suppliers center to provide reprocessing and mixed oxide fuel fabrication services to meet the demand of those advanced countries in West Europe already committed to plutonium recycle and the commercialization of breeder reactors. Our principal interest in this enterprise, however, would be to demonstrate that Third World countries would not be cut off from obtaining such services, at a later date, should they discover the need for mixed oxide fuel as a supplement to low-enriched uranium to fuel their reactors. Without prejudging the merits, time would be provided to consider whether centralized supplier services or multinational fuel centers, located outside of supplier territory, would best meet the needs of Third World countries, while providing the most effective system of international safeguards against the military diversion of plutonium separated from spent fuel.

The impact of the U.S. initiative would be enhanced if, over time, the other principal suppliers come to the same conclusion as we did. But its success would not depend on our securing an immediate consensus in this issue, with regard to their own programs. Rather, the success would depend on a developing consensus as to the wisdom of the U.S. decision, in terms of a common policy of restraint in transactions with Third World nations, and an agreement on steps other suppliers might take in their own "agreements for cooperation." Commissioner Gilinsky's remarks about steps the U.S. might take have an obvious bearing on potentially parallel initiatives by other suppliers, in addition to an agreement not to sell national processing plants:

> Let us turn to the specific restrictions that might be placed on plutonium to be generated in U.S.-supplied facilities and materials. *Ideally, we should like to see the spent fuel derived from these sources held unavailable for reprocessing until secure means can be provided for the use of the extracted plutonium.* [emphasis supplied] While the U.S. Agreements for Cooperation vary, they generally provide that all alteration and reprocessing of U.S.-supplied fuel must be done in facilities acceptable to both parties. In principle, these provisions might be employed to postpone reprocessing of this material. In that event, spent fuel could be stored in the importing country, lodged in international depositories, conceivably operated by the IAEA, or returned to the United States for storage. . . .

At the very least, the U.S. initiative would fortify the pursuit of a consensus in the exports of the principal suppliers that a high priority should be placed on the search for ways to insure that spent fuels not be reprocessed until genuinely effective international machinery can be provided to prevent the diversion of the extracted plutonium. In so doing, the U.S. would sharply put to them the question of nonproliferation priorities in the strongest terms. It would emphasize that the nonproliferation venture is much more likely to succeed if all the principal suppliers concentrate more effort and money on extending the life of the "front end" of the nuclear fuel cycle to meet their own needs, as well as the needs of the Third World, for additional sources of energy.

Americans are inadequately aware of the impact abroad of U.S. nuclear power decisions, particularly those made on their commercial merits. In those cases where they are accompanied by a nonproliferation proposal that coincides with the U.S. reading of the commercial realities, the influence is very great, in contrast to pious rhetoric, separated from the commercial realities.

The first reaction of the principal suppliers would probably be that the U.S. decision to defer reprocessing, even in its own case, was dead wrong, and certainly as it applies to the situation in their own countries, where they have already crossed the bridge on the road to reprocessing and plutonium recycle, and the pursuit of commercial breeder reactors. Despite this initial reaction, it would be difficult for the principal suppliers simply to ignore the U.S. decision to concentrate on the "front end" of the fuel cycle for several decades more, or for them to ignore the validity of the U.S. case for nonproliferation, in reaching this decision.

It is not inconceivable that several years from now, despite large European commitments already made, plutonium recycle will look less "inevitable" than it looks now, nor is it inconceivable that the various commercial breeder programs of the Europeans will have run into the same kind of problems that have plagued the U.S. ever since its own premature venture into the commercialization of breeders, dating back to 1955. Furthermore, active environmentalists are not limited to the United States, and the advanced countries are already encountering problems of technology and public acceptance of the plutonium cycle, involving large-scale reprocessing plants, breeders, and the storage of radioactive wastes.

While the success of the U.S. initiative does not depend on it, a consensus in favor of a controlled approach toward the transfer of enrichment and reprocessing technology is much more likely to develop if the other principal suppliers encounter similar problems with the "back end" of the nuclear fuel cycle. At that point, the reprocessing issue in the Third World could quickly start to diminish, and restrictions placed on the sale of reprocessing plants and efforts to develop multinational fuel centers, or centralized sources of supply, will be seen as they should be—not as efforts to retard the energy programs of Third World countries, but as efforts to contain and control the transfer of the most sensitive nuclear technology until genuinely adequate machinery of international control can be established.

At that point, the world might again begin to move down a joint road to keep the peaceful atom peaceful, and to provide a new source of energy with the convergence of commercial and nonproliferation objectives.

George B. Kistiakowsky

6

Nuclear Power:

How Much Is Too Much?

Introduction

I am not an expert on nuclear energy, or other power sources and their industrial exploitation, but a physical chemist with continuing interest—academic and industrial—in combustion and other high temperature phenomena, such as explosions. I was also involved to a lesser extent in some of the technical issues more relevant to this chapter. Before presenting the considerations which have influenced me as a technically trained generalist to become skeptical about the proposition of a rapid expansion of nuclear-steam-electric power generation, it seems appropriate to summarize my views on broader energy-related policy issues.

The central one, of course, is what should be the objective of American overall energy policy? If one believes in the urgency of achieving, as soon as possible, independence of the United States from the rest of the world in the matter of energy supply, then vigorous exploitation of all domestic energy sources—almost regardless of cost, hazards, environmental, and societal risks—becomes a *must*. But do we have to strive for such

GEORGE B. KISTIAKOWSKY, *Professor of Chemistry Emeritus, Harvard University, has won many honors in the world of science, including The Presidential Medal of Freedom, 1961. He was a member of the President's Advisory Committee, 1957–63, and also served as Special Assistant to President Eisenhower for science and technology. Dr. Kistiakowsky was chairman of the science board of ITEK Corporation for more than a decade.*

Dr. Kistiakowsky dissociates himself from the findings of any American Assembly meeting that has been or may be held on the subject of nuclear energy.

self-sufficiency? I do not believe that this is the right order of national priorities.

A sound choice of domestic actions in the energy field with a careful regard to their consequences should have the highest priority. In the near future we should accept the importation of oil and natural gas as unavoidable and treat the problem of preventing a punitive oil embargo as a major objective of our foreign policy.

In the following pages, I make some proposals for policies related to civilian nuclear power. These proposals are for the near future, that is, up to about the end of this century. It is obvious that if a nuclear war does not destroy industrial civilization, in the long run, fossil fuels will have to be conserved for petrochemical purposes and nuclear and solar power will become the primary sources of usable energy.

My consideration of the nuclear option and its alternatives is premised on the assumptions that first, the continuing importation of a substantial amount of our energy sources, such as petroleum and natural gas, will permit a more selective development of domestic sources, and second, that the cost of energy in constant dollars (as that of most other non-renewable natural resources) will continue to rise and that this, coupled with the rising chorus of warnings about environmental damage and impending scarcities, will gradually induce greater public acceptance of a variety of energy conservation measures. With conservation, the historic growth rate of total energy use of about 4 percent per annum, that made the United States into the greatest per capita user of energy, is certain to drop. Thus NEPLAN, the planning organization of the New England utilities has already revised its long-range electric power growth rate from 7.5 percent to 6.4 percent annually.[1]

Several aspects of nuclear power influence one's judgment about its desirability: its economics, the risks of catastrophic accidents, the disposal of massive amounts of radioactive byproducts, the safeguards from sabotage, theft, etc., with their social consequences, and of course, the availability and the disadvantages of other sources of useful energy. I shall consider these issues in the order indicated, not necessarily in the order of their importance.

On Economics

To a noneconomist from the middle classes one aspect of electric power is very obvious: lately, the cost of electricity has been rising faster

---

[1] *The Economics of Nuclear Power* (Boston: The Massachusetts State Energy Policy Office, 1975).

than income. This electricity is generated by a conglomerate of plants, new and old, some hydroelectric but mainly either coal-, oil-fired, or nuclear, whose outputs are combined for distribution by power grids now spanning much of the continent. What reaches the consumer is only about 30 percent of the heat energy released by the fuel, fossil or nuclear, and that figure is difficult to raise because of the Laws of Thermodynamics. What reaches the consumer is almost quantitatively convertible into useful work—or quantitatively into heat—but the remainder of the fuel energy, the 70 percent, is irretrievably lost as waste heat at the generating plant. Conversely almost all fuel energy can be locally utilized for heating purposes but petroleum fractions used in small engines (for example, autos) deliver less than 30 percent of their energy as work and the rest is rejected as waste heat.

In the early 1960s this writer, together with many others, listened to eloquent disquisitions of such enthusiasts as Alvin Weinberg of Oak Ridge about the great economies of large nuclear power plants and the resultant future of abundant cheap power. In New England in 1974 the electricity from fossil fueled plants (mostly oil) was twice as expensive as the average from its nuclear plants because of the high costs of imported fossil fuels. In the rest of the country the current costs of electric energy vary by more than a factor of two. Of the many figures reported by *Electrical World* (an industry periodical) for 1974, the lowest and the highest were from coal-fired plants, but the second highest and the third lowest were nuclear.[2]

These figures refer, of course, to plants whose construction began in the 1960s. Since then the capital costs have been rising rapidly, especially those of the nuclear plants. Hans Bethe [3] even now asserts that nuclear power will be cheaper in the future than alternatives, citing $730 as the capital cost per KWe. On the other end of the spectrum, a figure of $1,135 per KWe has been estimated for a nuclear plant to be initiated now and go on stream by about 1985, about 30 percent higher than the cost of a coal-fired plant of comparable size.[4] Other better documented sources give figures within these extremes.

The estimate of the relative future costs of electricity from coal-fired and nuclear plants is complicated. Since coal contains variable amounts of sulfur, and sulfur oxides are judged to be a major air pollutant, some utilities are being forced to install scrubbers of the flue gases (about

2 *Electrical World* (November 15, 1975).
3 Hans A. Bethe, "The Necessity of Nuclear Power," *Scientific American* (January 1976).
4 *New York Times* (November 16, 1975).

twenty-five are already in use and one hundred or so are in construction) which are quite expensive ($65 to $100 per KWe) and still have reliability problems. Oil-fired plants require no scrubbers but the present costs of oil make it the most expensive source of electricity.

The projections of future costs of coal, a far greater cost component than uranium fuel, vary depending on which authority one consults. Another uncertain cost item is the future capacity factor, that is, the fraction of design capacity that is actually in use. In recent years it was a disappointingly low average of 0.5 to 0.6 for nuclear plants.

On the basis of various such projections [5] the writer concludes that electricity from coal-fired plants, except possibly that in New England, is quite likely to be lower in the future than from nuclear plants. This comparison becomes even less favorable to nuclear power when certain hidden costs are included. The original development costs of small light-water reactors (LWRs) including the early demonstration power plants, the continuing R & D on nuclear construction materials by AEC and ERDA, as well as safety and waste disposal R & D, have all been charged off to the naval reactors and other government projects, but represent a not insignificant item to the taxpayers. Similarly the charge for the enrichment of the U-235 isotope in the fuel of commercial LWRs does not include the amortization costs of the existing (originally strictly military) gaseous diffusion plants. When additional enrichment plants are built—which will have to happen soon if nuclear power continues to expand—an expenditure of many billions of dollars will be required, which in one way or another will be charged to electric power users or perhaps the taxpayers generally. The quoted prices also do not include the costs of dismantling and permanent safe disposal of radioactively contaminated components of a reactor when it reaches the end of its useful life, perhaps in less than thirty years, judging by the practices with other high-pressure industrial equipment. Finally there is no allowance for very long-term safe disposal of spent fuel components. This, as well as the possibly costly provision of safeguards, will be considered below.

With all of this noted, the figures on the relative costs of nuclear and fossil fuel generation of electricity become rather fuzzy. They depend, it seems, to some extent on the ideology of the authors. The objective and incontrovertible fact is that the enthusiasm of the electric power industry for nuclear reactors has evaporated, at least for the time being. Thus the Energy Research and Development Administration (ERDA), the prime nuclear promoter in Washington, revealed on January 22,

[5] I. C. Bupp et al., "The Economics of Nuclear Power" *Technology Review* (February 1975).

1976 [6] that power companies announced the plans for only eleven new reactors in 1975 and have cancelled orders for thirteen others.

On Safety

Long ago (1950-60), as the Special Assistant to the President for Science and Technology, I represented the White House in the meetings of the Federal Radiation Council, constituted in response to public concern about the health effects of nuclear radiation. The Council's first task was to produce a government-wide set of safe radiation exposure standards—not a trivial task—partly because of the failure of the AEC to sponsor adequate earlier research on the subject. I mention these bygone events because the AEC Commissioner who was a member of the Council took a firm position—maintained throughout our sessions—that the standards must be high enough not to interfere with the AEC tasks. This was at a time when the uranium miners were being exposed to radiation to such an extent that mortality from lung cancer among them became several times higher than in the general population (this includes cigarette smokers).[7] A similar assessment of relative priorities seems to have characterized the subsequent actions of AEC while promoting nuclear power. Thus a sleek "Report to the President: Civilian Nuclear Power" transmitted by Glenn Seaborg, AEC chairman late in 1962, does not once mention problems arising from exposure to radiation and does not list safety among its "Objectives for the Future" or in a subsequent expanded summary of objectives. Near the very end of the text there is more than half a page devoted to "Reactor Safety: Siting Problems."

In 1966, after the first industrial orders for the very large reactors, AEC set up a study to consider their safety. In 1967 this task force recommended more intensive research on the adequacy of the planned Emergency Core Cooling Systems (ECCS) for the reactors. (The purpose of the ECCS is to inject cooling water into the core if the regular cooling by circulated water fails. Unless it is cooled, the highly radioactive core starts to overheat and, in about an hour, melts with possibly disastrous consequences.) Four years later experiments were made at last with "semiscale" (actually miniature) simulation apparatus. These experiments caused great surprise because the ECCS failed. This was contrary to predictions based on computer calculations like those on which the ECCS plant designs were based. The AEC then issued revised require-

6 *New York Times* (January 23, 1976).

7 H. W. Kendall et al., *The Nuclear Fuel Cycle* (Cambridge: M.I.T. Press).

ments for ECCS designs. These "interim criteria" were challenged by several groups including the Sierra Club, and the Union of Concerned Scientists. After prolonged "Rule Making Hearings,"[8] new, and substantially further tightened criteria for ECCS were issued by AEC, at the end of 1973. Meanwhile the AEC seemed to have decided to get rid of its tormentors once and for all and so instituted a multi-million dollar, largely in-house study to calculate, with the aid of computers, the probability of all sorts of accidents with the light-water reactor (LWR) plants. It issued in 1974 the Rasmussen Report: Draft WASH-1400 [9] which after careful study was criticized by a special study group of the American Physical Society [10] and by the Union of Concerned Scientists.[11] The final version of the Rasmussen Report has been recently released by the NRC (Nuclear Regulatory Commission), the relabeled AEC regulatory apparatus. The final version of WASH-1400 admits that casualties, prompt and delayed, will be in an order of magnitude greater in a statistically average catastrophic accident than the figures given in the draft. These earlier numbers were severely criticized by the APS and UCS as incomplete. The final version, however, persists in its view that its methods yield conservative calculations of the absolute probability of all catastrophic accidents and that this probability is almost vanishingly small.

I must disqualify myself as an independent referee of WASH-1400, but the portions of the debate that I understand force me to side with its critics who assert that WASH-1400 has not made allowances for *all* possible accident sequences; that this is virtually impossible to do with a system as complex and as poorly understood as a malfunctioning reactor; and that the report is especially weak in its treatment of the "common mode failure" of redundant subsystems. The latter criticism has been strikingly vindicated by the near-catastrophic accident at Browns Ferry in March 1975, where a lighted candle inserted into a cable tunnel led through a complex chain of events to a complete loss of the controls of one reactor, including all its ECCS controls, a temporary loss of its coolant water flow, and major damage to the controls including loss of ECCS of the other reactor.[12] Nearly a year later these reactors were still out of commission and were being repaired.

[8] J. Primak and F. V. Hippel, *Advice and Dissent* (Basic Books, 1974).
[9] *Draft WASH-1400* (US Atomic Energy Commission, 1974).
[10] *Reviews of Modern Physics,* 47, 1 (Summer 1975).
[11] H. W. Kendall and S. Moglewer, *Preliminary Review of the AEC Reactor Safety Study* (Cambridge: Union of Concerned Scientists, November 1974).
[12] *Wall Street Journal* (March 25, 1975); "Browns Ferry Nuclear Plant Fire Hearings" (Joint Congressional Committee on Atomic Energy, September 16, 1975).

According to the calculation of WASH-1400, the probability of a core meltdown in a LWR is about one in twenty thousand per annum. The critics of the report suggest that the calculation gives the lower bound rather than the true probability of this event. Unfortunately the experience to date with large LWRs is totally inadequate to provide statistically meaningful information; the large LWRs have been in operation only for about 200 reactor-years.[13] In this sample there were no meltdowns but two temporary coolant losses at Browns Ferry and the Dresden II reactor (not caused by pipe ruptures) and numerous minor safety-related incidents. In his earlier article in *Scientific American Magazine,* which has been cited, Bethe refers to "nearly" 2,000 reactor-years of operation in arguing for their safety. This figure is totally misleading because the issue is not whether light-water cooled and moderated reactors are generally and inherently safe. What is at issue is the safety of very large LWRs built commercially after competitive bidding. To get the "nearly 2,000" figure, the older, small, power reactors and about 1,500 reactor-years of naval ship reactors were included. These are only about 0.05 of the size of modern electric power units, use a different core (nearly pure U-235), were built on a noncompetitive basis to exacting navy standards, and their safety records have not been made public (although it has been requested).

The WASH-1400 report concludes that about 99 percent of all meltdowns [14] will cause at most only moderate disasters because the molten reactor core after penetrating through the wall of the pressure vessel and the concrete flooring of the containment building (this is quite probable after the melting according to WASH-1400) will sink into the subsoil and become sealed there, with only the volatile fission products diffusing into the atmosphere, causing delayed cancers, perhaps, but no prompt fatalities. As an individual with some experience in violent phenomena, I find the latter part of this scenario quite optimistic rather than conservative. One would be dealing here with a couple of hundred tons of molten metal and ceramics, perhaps at about 5000° F and continuing to generate heat and, because of the radioactivity of the fission products, at a rate comparable to that of a furnace burning several hundred tons of coal per day. When this mass reaches the soil below the containment building with its ground water and thermally unstable components (for example, the carbonates), the event is likely to be very violent, resulting in a blow-out and the entrapment of fission products

13 The 1975 list of operating reactors. *Nuclear Engineering International.*
14 "Comments on Draft WASH-1400, Reactor Safety Study" (Washington D.C.: Environmental Protection Agency, November 1974).

and plutonium in the escaping vapors and dust, coming down as radio-active fall-out downwind. Having concluded this I am unable to revise quantitatively the predictions of the WASH-1400 and in fact doubt that it is possible without a great deal more test data than are available.

A strong indication that the electric power industry and the Washington staffs themselves do not quite believe the reassuring findings of WASH-1400 is the fierce lobbying pressure they put on Congress to pass the extension of the Price-Anderson Act late in 1975. This act protects the industry from great disasters by providing federal indemnification against damage suits above $120 million and limiting all liability of the nuclear power industry to $560 million. The Ford Foundation Energy Policy Project commented thus before the passage of the extension:

> It is our view that the nuclear power industry is now sufficiently mature for a revision of the Price-Anderson Act, so that the market price will reflect the potential social costs of nuclear power. If nuclear power is indeed as safe as AEC and the industry claim, the risks are small enough to be borne by the enterprises involved. If the utilities are unwilling to build new plants on certain sites, or to buy reactors of certain designs, without the shield of the Price-Anderson Act, then those locations and those plants are too risky to be built.[15]

Indeed, if the findings of the WASH-1400 report are accepted as gospel truth, the likelihood of accidents with liabilities in the billion dollar class is so immeasurably low that it should be acceptable even if commercial insurance is not available. The utilities could certainly work out consortium arrangements to divide the liability in an infinitesimally probable accident. Evidently that is not the way the industry views the calculations of the WASH-1400 report and hence its own security.

That the compensation afforded by the Price-Anderson Act may be grossly inadequate can be illustrated by an admittedly extreme case. Consider the Indian Point reactors, thirty miles upstream from New York City on the Hudson, which were in the news in early 1976 because Mr. R. D. Pollard on the staff of NRC who was in charge of monitoring the safety of this installation resigned,[16] charging that NRC suppressed the existence of unresolved safety problems and that the Indian Point Reactor #2, now operational, is unsafe. Within twenty miles of the Indian Point site live nearly a million people, mostly south of it, and the prevailing winds blow south toward New York City. The magnitude of a disaster resulting from an explosive failure of the pressure vessel and the containment building, and the release into the atmosphere of

---

[15] *A Time to Choose* (Cambridge: Ballinger Publishing Company, 1974), pp 222-23.
[16] *New York Times* (February 10, 1976).

massive quantities of fission products is hard to estimate. The probability of this happening is extremely low, but if it does happen the resulting monetary losses from prompt and delayed casualties, and the contamination of the land would run into many billions of dollars; yet either the victims or next of kin would be compensated by only pennies on the dollar.

What is the probability of the rupture of the pressure vessel? WASH-1400 predicts one in ten million years (10^{-7} reactor-year) but a study made in Sweden arrived at a considerably higher figure,[17] and a well-known metallurgist, Sir Alan Cottrell, the chief scientist of the British government, testified in writing that the reliability of pressure vessels of the size used in the 1,000 MWe reactors is doubtful.[18]

The WASH-1400 asserts that on the average there will be two fatalities per 100 reactor-years. This is a very low accident rate, even allowing that it is only for accidents related to reactor misbehavior and does not include what happens in the rest of the power generating station, or the radiation and other injuries to miners, to workers in the ore milling plants, in fuel reworking, etc. I suspect though that even the total would come out no worse than typical figures for other industries, although critics point out that in addition to the two "prompt" fatalities there will be hundreds of long delayed victims of cancer and genetic damage.

My personal difficulty in weighing the acceptability of nuclear power from the point of view of safety is that because of the consistent past record of the AEC and its allied industry, of which only a fragment has been cited here, I find it difficult to trust anything they and their spokesmen state on the matter of safety. AEC has been now replaced by its offspring ERDA and NRC, but this seems to have changed little. Thus the Vermont Yankee reactor (whose activation earlier was delayed a year because of safety problems) was temporarily shut down in February 1976 because it was discovered that its design was such that a relatively minor malfunction might lead to a catastrophic failure. And now three senior technical staff members of the GE Nuclear Division have resigned [19] asserting publicly that the GE manufactured reactors, about half of those now operational in the U.S.A., are potentially subject to dangerous unrecognized malfunctions.

[17] J. P. Bento et al., "Comparison of WASH-1400 and the Swedish Urban Siting Study," Report A-483 (Sweden: Atomic Energy Company, 1975).

[18] Alan Cottrell, "Fracture of Steel Pressure Vessels," Cabinet Office, statement to the Select (Parliamentary) Committee on Science and Technology, January 22, 1974, London, England. The vessels in question are some fifteen feet in diameter, over thirty feet high, with walls up to twelve inches thick.

[19] *New York Times* (February 3, 1976).

The other complicating aspect of nuclear power acceptability is not technical but societal, arising from the nonuniform acceptability to the public of different sources and kinds of injuries. Thus our society resists measures that would reduce the numerous accidents due to the use of private automobiles or that would reduce fatal illnesses caused by cigarette smoking. On the other hand there is strong insistence on extreme safety of commercial aircraft travel or on the freedom of foodstuffs from suspected carcinogenic additives (dealt with by the Delaney Amendment to the Pure Food and Drug Act). There undoubtedly exist scholarly works explaining this wide intensity spectrum of public fears about the dangers to which it is exposed, but I am not familiar with them. Therefore it is only intuitively that I place nuclear power accidents in that part of the spectrum where the public fear is intense. To be socially acceptable, it seems to me, the American public will want nuclear power to be nearly fault-free in the matter of major accidents.

Numerous forced reactor shutdowns and reduced power operations [20] suggest that the nearly fault-free condition is far from being achieved. Thus nobody other than the Director of the WASH-1400 Reactor Safety Study, N. C. Rasmussen, wrote:

> Probably one of the most serious issues that the intervenors (critics of nuclear power) can raise today, with good statistics to back their case, is that the nuclear power plants have not performed with the degree of reliability we would expect from machines built with the care and attention to safety and reliability that we have so often claimed.[21]

On the Disposal of Radioactive Waste

The source of high-level radioactive waste is the spent fuel rods when they are chemically reworked to extract plutonium, and unspent uranium after a preliminary underwater storage at the reactor for a few months to let the radioactivity subside to a low percentage of the original intensity. In the reworking process several kilos of fission products end up as waste for each kilo of plutonium recovered.

What is ideally required is a mode of storage which will shield the uninformed humans of many generations to come from the waste and its radiation because nuclear radiation does its damage without any warning. The storage must be such that radioactive constituents will stay put as long as they are hazardous. These constituents must also be in such a

---

[20] D. D. Comey, "Will Idle Capacity Kill Nuclear Power?" *Bulletin of Atomic Scientists,* (November 1974).

[21] N. C. Rasmussen, *Combustion* (June 1974).

form that they cannot be extracted by living organisms of all varieties because some materials once absorbed or ingested concentrate to an extraordinary degree in the so-called food chain.

The duration of this storage need be no longer than some centuries for the high-level radioactive waste from fuel elements, after a more complete extraction of plutonium and other heavy artificial elements (the actinides) than has been practiced by AEC. If the long-lived plutonium and actinides are present, the isolation storage must last nearly geologic spans of time unless potential radiation damage to some humans in the far-off future is discounted as of no concern of ours. In 1972, Dr. Frank Pittman, director of the AEC Division of Management and Transportation, said: "While none of the suggested long-term solutions to the problem of permanent disposal of high-level radioactive waste is technically or economically feasible today, the AEC recognizes that one or more may well offer attractive options in the future." [22] Since then the criticisms of the nuclear power industry have greatly intensified, and in response its spokesmen have become highly reassuring. Thus, Bethe, in his article cited earlier said: "It seems virtually certain that a suitable permanent storage site will be found. It is regrettable that ERDA is so slow. . . ."

It has been the custom of AEC to bury contaminated scrap from nuclear installations and to pour lower level liquid wastes into trenches on guarded reservations because these are supposed to become more or less harmless within the human lifetime. The Maxey Flats disposal facility in Kentucky received over the years much of the industrial waste; altogether more than a million curies of radioactivity, and over eighty kilograms of plutonium. Recent investigation by the Environmental Protection Agency [23] revealed that plutonium has migrated in this humid forest site as far as a kilometer.

Studies of the soil under and around a leaky underground tank in the Hanford reservation (Richland, Washington), which leaked more than 100,000 gallons of high-level waste, did show the radioactivity to have penetrated at least eighty feet into the ground below the tank bottom and ninety feet laterally.[24] These are not unique instances, but Bethe has chosen for his *Scientific American* article the example of a rock site in the Gabon Republic where nearly two billion years ago a natural

[22] F. K. Pittman, "Management of Commercial High-Level Radioactive Waste," paper presented at M.I.T. (Summer 1972).

[23] G. L. Meyer, report at the Meeting of the International Atomic Energy Agency, San Francisco (November 17–21, 1975).

[24] *Atomic Energy Clearing House,* 19, 25 (June 18, 1973).

nuclear chain reaction seems to have taken place in a rich uranium ore deposit. According to Bethe the plutonium that was formed "did not move as much as a millimeter during its 25,000-year lifetime." A facetious reader will conclude that blind nature was wiser than the humans in charge of waste disposal. At any rate it is clear that the human problem is not solved, although its solution may be found some time in the future. The management of ERDA and NRC have indeed not decided on that permanent disposal site which Pittman wrote about in 1972.

The plutonium extraction for military purposes has been practiced at the Hanford, Washington site since 1944 and at the Savannah River, South Carolina site since about the mid-1950s. The many millions of gallons of the corrosive and heat generating waste slurry have been stored in underground tanks, pending final disposition. This operation has not been an unqualified success, as by the end of 1974 more than twenty leaks had been discovered in the tanks (and others in the associated piping), resulting in the loss of a half million gallons of sludge and solution into the ground.[25] Fortunately the ground water where the Hanford tanks are located is about 200 feet below surface and it has not thus far been contaminated. The leaks at the Savannah River site have been fewer and smaller.

Quite a few permanent disposal plans have been suggested from time to time, ranging from the placement of the waste in solar orbits by large rocket vehicles [26] to burial in Antarctic ice [27] and deep underground caverns created by the detonation of nuclear devices.[28] Temporary but centuries-long storage of solidified waste in "engineered surface facilities" [29] had been decided by AEC. The presently favored disposal involves converting the waste, after a temporary storage, into a glass-like, fused ceramic solid in the form of long cylinders weighing tons apiece. These, after being encased in steel jackets, would be stored in ventilated and guarded vaults and years later when the heat generation becomes sufficiently low, would be inserted into holes drilled in the walls of specially dug tunnels in a large rock salt deposit deep underground. The conversion into ceramic cylinders would involve some thousands of tons of solid waste annually by the year 2000 if ERDA's estimates of

[25] Report RED-75-309 (General Accounting Office, December 18, 1974).

[26] A. M. Weinberg, "Social Institutions and Nuclear Energy," *Science,* 177, 27, 1972.

[27] B. Philbert, *Schweizerische Zeitung fuer Hydrologie* 1, 262(1961); also J. O. Blomeke et al., *Physics Today* 26, 8 (1973), p. 36.

[28] US-AEC, WASH-1250 (December 1972); B. L. Cohen et al., *Nuclear Technology* (April 1972).

[29] W. W. Hambleton, *Technology Review* (March/April 1972), p. 15.

the expansion of the nuclear electric power industry, that is, over 500 large reactors in the U.S.A., indeed materialize.

This disposal scheme appeals to me as sounder than the other proposals because fused ceramics can be extremely resistant to all sorts of leaching and the salt beds appear to offer geologically stable long-term disposal sites. However, if the storage above ground is to be protracted, one must have greater faith in the stability of our civilization than is perhaps justified by past human history. Whatever the cause, ERDA has now suspended the AEC plans for vault construction on the Nevada AEC testing grounds. The original AEC plan to use salt deposits for waste disposal ended rather embarrassingly when AEC discovered, subsequent to requesting twenty-five million dollars from the Congress for the project, that next to the chosen location at Lyons, Kansas, the American Salt Company was engaged in large-scale salt mining by solution (that could flood the site some day), and old gas bore-holes were found to penetrate the chosen site.

Since the disposal of radioactive wastes could affect future generations unfavorably, it is to be hoped that before commitments are made there will be extensive test documentation available of which such papers as that of Blomeke [30] are a start, followed by the adversary-type extensive hearings open to the public.

The status of waste disposal in the other countries with nuclear reactors, most of which do not have the salt deposits of the U.S.A. or the Nevada proving grounds, is not clear to me. The British, after dumping containerized waste into the oceans for a time, stopped doing this possibly because they learned how the biosphere concentrates some radioactive elements.

It would seem that Sweden's decision in 1975 to suspend its policy of rapid expansion of nuclear power [31] was influenced by the problems of waste disposal, since Sweden has no oil but rich uranium deposits and a highly developed nuclear technology.

Currently in the United States no processing of fuel is being done and all spent elements remain in the water pools at the reactors. In 1966 an operating permit was issued by AEC to the Nuclear Fuels Services, Inc. for a reprocessing plant in West Valley, New York. Through 1971 the plant processed some 500 tons of fuel (only about twenty annual

[30] J. O. Blomeke, "Management of Wastes, etc.," proceedings of the Joint Topical Meeting of Commercial Nuclear Fuel Technology Today. Toronto, Canada (April 1975), sponsored by American Nuclear Society, etc.

[31] Uno Svedlin, "Sweden's Energy Debate," *Energy Policy* (September 1975), p. 258.

reactor discharges), was then shut down for renovation and expansion and has not been reopened. Judging from the reports of the Department of Environmental Conservation of the State of New York and other sources,[32] somewhat greater than wholly acceptable quantities of volatile radioelements escaped into the atmosphere and the others escaped into the creek flowing near the plant site. Thus a more elaborate containment will be required from the full-scale fuel reprocessing plants of the future.

This section would not be complete without commenting on the harmful biological effects of nuclear radiation under circumstances other than waste disposal or massive accidents. After fighting the bitter rearguard battles to keep the radiation exposure standards high for many years, the AEC and now the NRC have been forced to lower the permissible radiation levels from reactors and in mines way down and so have made peace with most of their critics.[33] Disputes still rage about the toxicity of plutonium reaching the environment in one way or another since it is apparently a very toxic substance when breathed in as dust. The problem of radon emanation and radioactive stream pollution from the tailings of uranium mines is another significant and locally serious problem.[34]

I find myself unable to contribute meaningfully to the resolution of these medico-environmental problems, which I believe are not central to the future of nuclear power. With adequate public pressure they can be resolved—albeit at considerable cost—in such a way that radiation injuries will be only a very minor contribution to the general toll of modern technology and, therefore, be acceptable to society.

On Safeguards

Our times, unfortunately, are not an era characterized by the longevity of social institutions which Weinberg[35] considers important to a safe growth of civilian nuclear power industry. Ours is an era of wars, of revolutionary upheavals and military takeovers. Domestically our society is troubled by the frequency of violent crime, by acts of sabotage and terrorist activities. Not only handguns but automatic weapons and high explosives seem to be readily available to the disaffected.[36]

[32] P. M. Hatfield, *The Nuclear Fuel Cycle* (Cambridge: The M.I.T. Press, 1975).

[33] See for instance "The Effects on Population of Exposure to Low Levels of Ionizing Radiation," NAS-NRC, Washington D.C., 1972.

[34] H. P. Metzger, *The Atomic Establishment* (New York: Simon and Schuster, Inc., 1972).

[35] A. M. Weinberg, "Social Institutions and Nuclear Energy," *Science*, 177, 27, 1972.

[36] *The Threat to Licensed Nuclear Facilities*, technical report MTR-7022, MITRE Corp. (1975).

It is against this background that one must give consideration to the safeguards of nuclear installations against sabotage, terrorist takeovers, theft, and vehicle highjacking. Such an inquiry must also encompass the world beyond the borders of the United States because the expansion of domestic nuclear power, the export of this technology, and the proliferation of nuclear-armed states are not unrelated phenomena.

The threats to civilian nuclear power installations should not be exaggerated, however. There have been no publicly recorded, aggressive incidents in the past, even though plutonium and weapons-grade U-235 have been produced in several AEC installations since 1944. Many tons of these "special nuclear materials" have been transported to fabrication facilities such as Rocky Flats in Colorado, and then the fabricated items have been transported by the thousands to the military storage sites. This has been accomplished at the price of maintaining the activity as an expensive, essentially military operation [37] with tight security, including elaborate personal security checks required for the "Q" type clearances of the more sophisticated personnel. The personnel have to surrender some of their civil rights; for instance, submitting to the lie detector test at the Kerr-McGee (Cimarron, Oklahoma) fuel processing plant as a condition of their employment. [38]

The present activities of the American civilian nuclear power industry involve no weapons-grade fissionable materials since the uranium fuel is enriched only to about 3 percent of U-235, and the spent fuel is stored under water containing plutonium in a highly radioactive medium. Thus the possible threats are random sabotage and the terrorist takeovers with the objective of blackmail by threats of catastrophic fall-out release through plant damage. The MITRE report is uninformative about the specifics of these problems. The report of the Massachusetts Commission on Nuclear Safety is more helpful. It states:

> In the absence of adequate security protection, a very small number of knowledgeable people could bring about a "melt-down" in a nuclear power plant and cause a breach of the containment with a consequent release of radioactivity to the environment. Furthermore, they could select a time when meteorological conditions were such as to produce maximum damage. Thus, what might be a maximum credible, but extremely unlikely, accident on the scale mentioned above could be perhaps one of the more likely consequences

---

[37] The ERDA FY 1977 budget request for weapons materials and production is $984 million ($445 + $538), although the quantities of fissionable materials involved are quite small compared to those required to keep going at the normal loads, say, 200 reactors of 1,000 MWe capacity each.

[38] *New York Times* (January 7, 1975), p. 14.

of sabotage, *assuming* determined and knowledgeable terrorists. The scale of damage, the extreme public sensitivity to the idea of nuclear radiation, and the vulnerability of a nuclear reactor are such that a reactor would, in the absence of strong security measures, be one of the more attractive targets for any terrorist group.[39]

The report goes on to observe that the existing plant protection is not commensurate with the threat and that two explosives technicians, after visiting a nuclear power plant, told the Commission that they would have no trouble disrupting operations there if so inclined. From my own experience as an ex-explosives expert, and after viewing photos of the plants, etc., I would tend to agree with these individuals. The disruption by a determined armed group is made much easier, of course, by the availability of blueprints of the nuclear power plants for public inspection as required by law.

How imminent the real threat is, I am unable to judge. To eliminate it beyond reasonable doubt would require, for each installation, not only the usual security fence, guard at the gate, and employee passes, but a rather substantial force of well-armed guards, perhaps six on each shift, and somewhat intrusive monitoring of the employees to insure that none become accomplices to the sabotage attempt. Clearly the social climate in such an installation would be different from that in other public utility plants and this might be justified only in selected places near large population centers, such as Indian Point, New York or Pilgrim, Massachusetts.

So long as the spent fuel rods are not reworked chemically, the civilian nuclear industry will not produce materials suitable for weapons purposes and the problem of theft does not arise. However, the proponents of a rapid growth of the industry insist that the present state can only be very temporary because of limited supplies of uranium. Indeed only about 0.4 percent of total uranium—that is, about half of the U-235 content—is subjected to fission presently. Reworking the spent fuel elements, putting the extracted uranium, which is richer in U-235 than the natural material, back into the enrichment process and putting the extracted plutonium into the new fuel elements would more than double the quantity available for fission. Finally, when (and if) the breeder reactors take over, virtually 100 percent of uranium will become available for electric energy production.

A report of the Stockholm International Peace Research Institute [40]

[39] Massachusetts Department of Public Health, Boston, September 1975.

[40] Stockholm International Peace Research Institute, *Nuclear Proliferation Problems* (Stockholm: Almquist and Wiksell Company, 1974).

estimates that the American uranium reserves known in 1971, which would cost less than fifteen dollars per pound to extract, was about 450,000 tons and additional rich reserves were estimated to be over a million tons. A recent ERDA report [41] estimates the total potential domestic uranium resources, in deposits concentrated enough that the cost of recovery would be no more than thirty dollars per pound, as 3.6 million tons of uranium. The indicated cost, called the "forward cost," does not include the costs of exploration, the capital, or the profits. However the total cost—perhaps double the above—is still very much lower than that of fossil fuels, even if as at present, only 0.4 percent of uranium is utilized.

Approximately 500 tons of natural uranium are needed to start a 1,000 MWe LWR-type reactor, and 150 tons per year are needed to keep it operating. These figures tell us that unless the nuclear power industry expands at a rate leading to 500 or 1,000 reactors by the year 2000, there exists no urgency to start fuel reprocessing for quite a few years; after all, a nuclear power plant now takes nearly ten years from start to operation. By the same reasoning operational breeder reactors could be delayed into the twenty-first century.

The installed capacity of nuclear reactors in the United States is larger than in the rest of the world put together; [42] noncommunist world resources of uranium are at least double those in this country, and there appear to be large deposits in the communist world. Thus, the same lack of urgency to start fuel reprocessing would seem to prevail world-wide, unless one takes seriously the extraordinarily optimistic expansion plans of some less developed countries that seem not to take into account the capital requirements of nuclear installations.

Unfortunately the doubling of the available fissionable material by the reprocessing of spent fuel elements means a long step toward the plutonium economy with its special problems. A 1,000 MWe LWR produces something like 200 kilograms of plutonium per year, and hence in the steady state, nearly that much would go back into fresh fuel; this would be repeated in hundreds of reactors. A crude implosion bomb needs something like ten kilograms of plutonium and it is the contrast of these numbers which legitimately alarms all but the committed enthusiasts of nuclear power.

Extracting, annually, many tons of plutonium from spent fuel, transporting it to fuel fabrication facilities, handling it there, and transport-

---

[41] ERDA-33, Nuclear Fuel Cycle (March 1975).
[42] *Handbook on Nuclear Proliferation,* U.S. Congress, printed for the Committee on Government Operations (December 1975).

ing fuel to reactors will present many opportunities for diversion, theft and highjacking of a material which is far more valuable per pound than gold now. Even the unused fuel rods containing a couple of percent of plutonium would be a tempting object. The separation of plutonium from uranium is described in open literature and to accomplish it would not require chemistry Ph.Ds and sophisticated chemical plant facilities. When and if the breeder reactors of a type resembling the proposed Clinch River demonstration plant become operative, the difficulties will increase greatly because much more plutonium will be handled and the fuel rods will contain it in a highly concentrated state.

The objective of diversion of plutonium would be either domestic acts of blackmail and terrorism or clandestine export. In my estimation none of the past American terrorist activities involved groups with adequate technical sophistication to purify plutonium and to make even a very crude atom bomb.[43] To start with, they would not know how to get the necessary information from the open technical literature. But one cannot guarantee that technically trained individuals will not become involved in the future. In any case, what matters is not my assessment but the public perception of the possibility of acts of terrorism, threatening an explosion thousands of times more violent than the one in early 1976 of twenty-five pounds of dynamite in the LaGuardia airport and one that would spread many millions of *theoretically* "lethal" doses of plutonium oxide in the neighborhood. It seems to me that the public reaction to these possibilities would be rather intense. To neutralize it, quite elaborate security measures are most likely to be introduced by the nuclear industry: personal security investigations of all personnel counted in very many thousands; many thousands of armed guards in hundreds of installations throughout the country; armed fuel convoys by the many hundreds (six truckloads per plant annually [44]) crisscrossing the country. To this would be added the unavoidable active measures of security: telephone taps; covert surveillance; infiltration of "subversive" groups; etc. The impact on civil liberties of the populace could easily be a major one,[45] comparable to a perpetual state of war. Thus the FBI Director Kelly [46] has already stated that "we may have to surrender some civil liberties" to combat growing terrorism in general.

43 M. Willrich and T. B. Taylor, *Nuclear Theft: Risks and Safeguards* (Cambridge: Ballinger Publishing Company, 1974).

44 "Nuclear Power and Environment," American Nuclear Society (September 1974), p. 24.

45 "Policing Plutonium: The Civil Liberties Fallout," *Harvard Civil Liberties Law Review* 10, 2 (Spring 1975).

46 (John Chancellor), NBC newcast (January 13, 1976).

Add to this the indicated consequences of the plutonium economy, and some irreparable damage may be inflicted on our democracy, even if there is no theft followed by a country-wide search, or a terroristic act followed by mass hysteria.

Diversion by theft or highjacking of plutonium for foreign customers appears to me to be a more realistic and immediate threat. Terrorist activities in some troubled parts of the world are now on such a scale that the purchase of plutonium on the black market and its weaponeering utilization appear to be quite realistic consequences of the availability of plutonium in the civilian economy. Thus organized crime in the United States may consider the export of plutonium a useful addition to its import of narcotics. Higher technology devices than trained dogs will be necessary to sniff out plutonium in airports, docks, and border guard houses.

My last consideration is a speculation in the domain of geopolitics for which I obviously lack competence, but the issue is so grave that I cannot bypass it here. The policies of the superpowers, for instance the breach of their commitments in the Nonproliferation of Nuclear Weapons Treaty of 1968 to engage in steps of nuclear arms reduction, and the continuing nuclear arms race, have impressed the rest of the world with the overwhelming importance of nuclear weapons as an instrument of international power and influence. And so Great Britain, France, China, and India have reached the end of the road to nuclear explosives, one or two other states are estimated to be very near it, and a number have publicly stated their interest in doing the same.

The really difficult and costly part of getting fairly primitive but militarily usable atom bombs is the acquisition of fissionable material. If that is available, a multi-disciplinary group of twenty to thirty individuals with scientific and engineering training plus technicians could produce an acceptable weapon in no more than a couple of years without any great industrial backup, since so much information is available in the open liteurature.[47]

The United States is preeminent in the realm of the civilian nuclear power industry—it has nearly as many power reactors operating and under construction as the rest of the world put together. If the United States proceeds with the rapid expansion of this industry domestically, starts fuel reprocessing, plutonium utilization, and then introduces breeder reactors, it is difficult to see how the rest of the world could be induced not to follow. According to the International Atomic Energy

[47] Reference 46 estimates $18 million to design and produce 10 weapons of 20 KT yield over 10 years.

Agency (IAEA), some seventeen states are planning to have reprocessing capability five years hence. The inevitable consequence of these trends will be dozens of states, large and small, in the possession of stocks of atom bombs and the means of their delivery (thanks to irresponsible sales of modern arms to all who can pay for them and some that cannot). The accretion of nuclear weapons can happen so soon that the current political tensions will not have subsided in the world. Local wars are likely to continue flaring up and some day one of them will start the use of nuclear weapons, perhaps as an act of national desperation. Several times since the end of World War II local conflicts have drawn in more powerful states, and when that happens in a local nuclear war the amplification of the conflict could easily be nuclear also.

At present the threat of nuclear war is far less serious. Large uranium enrichment facilities are possessed only by the United States and the Soviet Union. France, Great Britain, and China have small enrichment plants needed for their weapons production but not large enough to feed a large domestic nuclear power industry and provide for export trade.

The Soviet Union's export of large power reactors appears to be limited to the Warsaw Pact nations and the nuclear export trade of the United States is much larger than the total of its competitors—France, Germany, Great Britain, Canada, and Japan. Thus, over-all, the United States can have a virtually dominant voice in controlling the rate of expansion of civilian nuclear industry in the noncommunist world. Unfortunately it does not exercise it, if one believes the story [48] that the agreement of exporting countries reached in the fall of 1975 permits the export of reprocessing facilities at the insistence of France and West Germany.

On Alternatives

Summing up, I cannot but conclude that the unrestricted growth of the civilian nuclear electric power industry involves a great many uncertainties and dangers, some of unpredictable and uncontrollable character. But are there alternatives to it that would permit the economic growth of our country? I believe there are some that would be effective at least in the near future, but in view of space limitations I can only mention them briefly.

To comment on the alternatives one must have some perspective on

[48] *New York Times* (February 24, 1976).

nuclear power as a present and future part of the total energy picture. The writings of some enthusiasts leave the impression that the growth of nuclear-generated electricity will resolve the entire energy crisis. Actually in 1973 in the United States, about 56 Q units of energy [49] were used as fuel and 6.4 Qs went to consumers as electricity. Producing this electricity took 9 Qs of coal, 7 Qs of oil and gas, 1 Q of nuclear energy (more than two-thirds of all these resulted in waste heat), and 3 Qs of water power. The largest use of petroleum, nearly 19 Qs out of a total of 34 Qs, was used for mobile power sources—autos, trucks, agricultural machinery, buses, locomotives, planes, and ships. Gas was mainly used in various heating applications—industrial, commercial, and residential. Both oil and gas are indispensable for the petrochemical industry, including the production of plastics and cheap fertilizers.

Few of these applications are likely to switch to nuclear generated steam or electricity in the next few decades. There may be some nuclear ship propulsion and some industrial heat where concerns about reactor safety are minimal—perhaps more if reactors become relatively cheap again—but the prospects of cheaper construction costs are not bright. Thus the main near-term contribution of expanding nuclear power to the resolution of the energy crisis would be a significant but by no means decisive reduction in the rising demand for oil and gas in the production of electricity. This production has been rising faster than our total use of energy but limits to this relative growth are in sight.

Since the depletion of domestic oil and gas are forecast [50] to come very soon after—if not before—the end of this century and the world reserves are also limited, the increased use of these most versatile energy sources is not a viable alternative to nuclear power.

The solution requires a concerted exploitation of several options, none rich enough to solve the problem alone. The most immediate and very significant is energy conservation, that is, the reduction of waste and the conversion to more efficient devices. The particulars of conservation actions, what they would entail, and their returns are so thoroughly and well presented in the Ford Foundation report that I need not dwell on this subject, except to emphasize its importance and to note that public education and political measures are central to progress in this direction.

[49] A Q is equal to 10^{15} BTUs, and 1,000 BTUs are equal to 0.3 of a kilowatt-hour of electricity.
[50] Lou C. Ruedsili and M. W. Firebaugh, eds., *Perspectives on Energy* (New York: Oxford University Press, 1975).

An equally important option is the exploitation of our huge coal reserves including those in western states. That means a great deal more stripmining, but the Federal Republic of Germany has shown that the countryside can be fully restored after stripmining and the same must be required by federal law in the United States, although it means increased price of coal. Assertions have been made that lack of water in the West makes restoration impossible. But California brings water over hundreds of miles and through the mountains to be used in irrigation. Why not have a federal pipeline bringing water from the Missouri-Mississippi basin to Wyoming, Dakota, and Montana to help restore stripmined lands?

Burning coal pollutes the atmosphere but modern technology can certainly cope with this blight, presently by scrubbers, and in the future perhaps by more sophisticated combustion processes being developed in Europe.

Increased exploitation of coal is also essential for making synthetic gas a substitute for the dwindling natural resources. This technology is in need of stronger federal support and leadership than the nuclear-committed ERDA is providing now; given that support, success is virtually certain because technology is well advanced.[51] On the other hand, coal hydrogenation into liquid fuels and their extraction from oil shale, which must come eventually unless our entire transportation and living mode are to be scrapped, are too far in the future to be significant options in this century.

Dismissed as trivial by nuclear enthusiasts are several sources which together can provide energy equivalent to perhaps a decade's growth in American demand. Three of these are already in minor use: partial solar heating of residential and industrial buildings and residential water tanks by fixed roof panels, electricity from sophisticated windmills which could be a significant contribution in the windy part of the country that was known as the dustbowl in the 1930s, and the geothermal sources that already provide half the electricity for San Francisco.

The point of these remarks is that there exist ways to keep down the consumption of oil for the next twenty years or so without expanding nuclear power or suffering recessions. What is needed to accomplish this is an enlightened federal government and the willingness of our people to move in that direction. Do I need to add that these are not minor problems?

51 "Energy R & D and National Progress," Office of Science and Technology, (U.S. Government Printing Office, 1964).

Conclusion

I find that the technology is not ready for a massive expansion of nuclear power and that our society is not ready to live with it. Nor is the world ready to live with it. I am in favor, therefore, of proceeding domestically with the alternatives whose drawbacks, by no means minor, can be foreseen and controlled. Those nuclear power reactors which are now operational, I would allow to continue operations and those under construction to be put on stream, but with all of them under more effective safety controls than those that seem to be practiced now.

As regards further additions, they should be few and limited to the sites quite far from population centers. Thus the added costs of power transmission would probably assure a modest rate of nuclear power expansion without the need for limitation by governmental action. Certainly NRC should apply strong pressure through its licensing power for more complete standardization of plant design than exists now, thus making possible more thorough elimination of potential sources of malfunction.

Fuel reprocessing should be held in abeyance and spent fuel rods be kept in underwater storage, pending a far better understanding and solution of the problems discussed above—and probably some others of which I am not aware.

The Clinch River demonstration breeder reactor plant should be postponed and the money put to better use. If the information which I gained from daily press reports is valid, this plant would have 350 MWe capacity and the current cost estimate is $1.9 billion. That means $5,500 per KWe. Some demonstration! Bupp and Derian [52] point out that there exists a simple relation between the acceptable cost differential of LWR and breeder reactors and the price of uranium fuel. Their analysis suggests that at the current prices the breeders would have to cost well below $1,500 per KWe to make sense economically. Since the proposed moderation in the growth of nuclear power would postpone the *need* of production breeders for several decades, a delay of several years in the start of the Clinch River plant is certainly reasonable.

The export of nuclear technology should be an important facet of the foreign policy of the United States since it is coupled with the proliferation of nuclear weapons. To chart the right course here is a precar-

52 I. C. Bupp and J. C. Derian, "The Breeder Reactor in the USA: A New Economic Analysis," *Technology Review,* 76, 8 (1974), M.I.T.

ious undertaking for an outsider. Assuming, as I do, that the prevention or rather the postponement of weapons proliferation should be our major policy objective one has to face the fundamental technological fact that once a country possesses not only nuclear power reactors, but also an operational fuel reprocessing facility, the production of nuclear arms becomes a political decision. It can be accomplished within a couple of years with no strain to the economy of even a poor country. Reprocessing facilities are a cheaper and shorter route to atom bombs than an enrichment plant, especially if the plant is built for a low enrichment factor. Considering the frequency with which governments are changed these days and the ease with which foreign commitments are abrogated, membership in IAEA and a signature under the Nuclear Proliferation Treaty still leave the very high probability that acquisition of reprocessing facilities by a country would be shortly followed by nuclear arms.

I believe that the United States, while allowing American corporations to export power reactors, should forbid the export of reprocessing technology except to nations that already have it. The United States has nearly a stranglehold on enriched uranium fuel; furthermore I understand that our chief export competitors, France and West Germany, import from us certain essential reactor components which they sell abroad as parts of the LWRs built by them under licenses from the General Electric Company or Westinghouse. Thus the United States is in a position to insist on an understanding with its allies that reactors be sold abroad with the proviso that for a certain span of time, say ten years, the buyer country refrain from building fuel reprocessing facilities at the risk of losing further supplies of nuclear fuel, and that spent fuel be turned over to a designated international facility for temporary storage.

The result of putting this suggestion into effect would be more restrictive policies on the expansion of civilian nuclear power industry than what is practiced now. There are obviously difficult obstacles in the way of this change, but now the United States and its principal allies, for the sake of commercial gains, it seems, are in effect sponsoring the process of world-wide proliferation of nuclear weapons.

I have tried in this chapter to trace some of the important aspects of the complex issue of expanding civilian nuclear power and to suggest some policies for the future. For reasons of space I have had to be selective and can only hope that I have considered the key problems. Perhaps the central objective which motivated me was to seek ways to postpone the proliferation of nuclear weapons until the threats of war subside in this world of ours.

Index

The American Assembly
COLUMBIA UNIVERSITY

About The American Assembly

The American Assembly was established by Dwight D. Eisenhower at Columbia University in 1950. It holds nonpartisan meetings and publishes authoritative books to illuminate issues of United States policy.

An affiliate of Columbia, with offices in the Graduate School of Business, the Assembly is a national, educational institution incorporated in the State of New York.

The Assembly seeks to provide information, stimulate discussion, and evoke independent conclusions in matters of vital public interest.

AMERICAN ASSEMBLY SESSIONS

At least two national programs are initiated each year. Authorities are retained to write background papers presenting essential data and defining the main issues in each subject.

A group of men and women representing a broad range of experience, competence, and American leadership meet for several days to discuss the Assembly topic and consider alternatives for national policy.

All Assemblies follow the same procedure. The background papers are sent to participants in advance of the Assembly. The Assembly meets in small groups for four or five lengthy periods. All groups use the same agenda. At the close of these informal sessions participants adopt in plenary session a final report of findings and recommendations.

Regional, state, and local Assemblies are held following the national session at Arden House. Assemblies have also been held in England, Switzerland, Malaysia, Canada, the Caribbean, South America, Central America, the Philippines, and Japan. Over one hundred institutions have co-sponsored one or more Assemblies.

ARDEN HOUSE

Home of The American Assembly and scene of the national sessions is Arden House, which was given to Columbia University in 1950 by W. Averell Harriman. E. Roland Harriman joined his brother in contributing toward adaptation of the property for conference purposes. The buildings and surrounding land, known as the Harriman Campus of Columbia University, are fifty miles north of New York City.

Arden House is a distinguished conference center. It is self-supporting and operates throughout the year for use by organizations with educational objectives.

The background papers for each Assembly program are published in cloth and paperbound editions for use by individuals, libraries, businesses, public agencies, nongovernmental organizations, educational institutions, discussion and service groups. In this way the deliberations of Assembly sessions are continued and extended.

The subjects of Assembly programs to date are:

1951—United States-Western Europe Relationships
1952—Inflation
1953—Economic Security for Americans
1954—The United States' Stake in the United Nations
　　—The Federal Government Service
1955—United States Agriculture
　　—The Forty-Eight States
1956—The Representation of the United States Abroad
　　—The United States and the Far East
1957—International Stability and Progress
　　—Atoms for Power
1958—The United States and Africa
　　—United States Monetary Policy
1959—Wages, Prices, Profits, and Productivity
　　—The United States and Latin America
1960—The Federal Government and Higher Education
　　—The Secretary of State
　　—Goals for Americans
1961—Arms Control: Issues for the Public
　　—Outer Space: Prospects for Man and Society
1962—Automation and Technological Change
　　—Cultural Affairs and Foreign Relations
1963—The Population Dilemma
　　—The United States and the Middle East
1964—The United States and Canada
　　—The Congress and America's Future
1965—The Courts, the Public, and the Law Explosion
　　—The United States and Japan
1966—State Legislatures in American Politics
　　—A World of Nuclear Powers?
　　—The United States and the Philippines
　　—Challenges to Collective Bargaining

1967—The United States and Eastern Europe
 —Ombudsmen for American Government?
1968—Uses of the Seas
 —Law in a Changing America
 —Overcoming World Hunger
1969—Black Economic Development
 —The States and the Urban Crisis
1970—The Health of Americans
 —The United States and the Caribbean
1971—The Future of American Transportation
 —Public Workers and Public Unions
1972—The Future of Foundations
 —Prisoners in America
1973—The Worker and the Job
 —Choosing the President
1974—The Good Earth of America
 —On Understanding Art Museums
 —Global Companies
1975—Law and the American Future
 —Women and the American Economy
1976—Nuclear Energy
 —Manpower Goals
 —Capital Needs in America

Second Editions, Revised:

1962—The United States and the Far East
1963—The United States and Latin America
 —The United States and Africa
1964—United States Monetary Policy
1965—The Federal Government Service
 —The Representation of the United States Abroad
1968—Cultural Affairs and Foreign Relations
 —Outer Space: Prospects for Man and Society
1969—The Population Dilemma
1972—The Congress and America's Future
1975—The United States and Japan